CHITIN:
Fulfilling a Biomaterials Promise

CHITIN:
Fulfilling a Biomaterials Promise

EUGENE KHOR
(Ph.D., Virginia Tech.)

Department of Chemistry,
National University of Singapore,
Republic of Singapore

2001

ELSEVIER
Amsterdam – London – New York – Oxford – Paris – Shannon - Tokyo

ELSEVIER SCIENCE Ltd
The Boulevard, Langford Lane
Kidlington, Oxford OX5 1GB, UK

First edition 2001

Library of Congress Cataloging in Publication Data
A catalog record from the Library of Congress has been applied for.

British Library Cataloguing in Publication Data
A catalogue record from the British Library has been applied for.

ISBN: 0 08 044018 5

♾ The paper used in this publication meets the requirements of ANSI/NISO Z39.48-1992 (Permanence of Paper).
Printed in The Netherlands.

DEDICATION

To the memory of 2LT Edmund Piper

A Symbol of a generation of young men
whose sacrifice was instrumental in building a Nation called Singapore

PREFACE

The incessant search for alternative and better ways to treat bodily ills is filled with challenges and opportunities. When it concerns replacement of body parts with artificial substitutes, state-of-the art technology developments such as Tissue Engineering are fueling the quest for better biomaterials that can meet a myriad of challenges. Central to this issue is the potential for the utilization of materials from nature. Among the candidates, chitin has been poised to be one such natural material that can be the answer to a variety of needs in the biomedical field. Ask any chitin researcher about the benefits of chitin and you will receive an earful of its capabilities as wound dressings, in bone substitutes and as drug delivery carriers, conveyed in a manner that is almost magical. Such is the passion, yet more is the accompanying confusion and lack of consensus that confronts chitin as it strives to be a fully qualified member of the biomaterials club.

Why is the situation unclear as to whether chitin will blossom into a fully accepted medical material? The ever-looming impediment to the full-scale launch of chitin as a biomaterial remains the production of high purity grades of the material and the daunting requirements of regulatory approval. Chitin biomedical products have appeared, and there has been a noticeable effort towards the introduction of biomedically suitable chitin. Nonetheless, the widespread proliferation of chitin-based medical products has not really taken off. The key question now is whether chitin will be a contender that truly makes it! With so much information amassed from chitin research over the past 40 years, what more can and should be done with or for chitin? For all its purported therapeutic benefits, can chitin succeed in vying for a market share in niches already occupied by cellulose, collagen, hyaluronic acid and the up and coming chondroitin and keratan sulfates. Will chitin be sidelined by these other biopolymers or newer technologies on the horizon? Will chitin, having exploded with promise, find itself falling short of erupting with profits?

So far, little has been discussed on the groundwork necessary for chitin to claim its rightful place as a biomaterial. This book is written with the aim to underscore the factors that must increasingly transpire in standardizing chitin processing and characterization. It attempts to capture the essential interplay between chitin's assets and limitations as a biomaterial, placing the past promises of chitin in perspective, addressing its present realities and offers insight into what is required to realize chitin's destiny (that includes its derivative, chitosan) as a biomaterial of the 21st century. For both the industrialists and researchers with a vested interest in commercializing chitin, I hope this book will serve as a primer towards this goal.

As in any endeavor of this nature, there are many people to acknowledge. First and foremost, I am grateful to four men, without whom this book would never have been written. To Professor Larry T. Taylor, my model of a complete scientist, for preparing me well for the scientific challenges in my life. To Professor KY Sim, my former Head of Department, who made it possible for me to learn from two distinguished Scientists: Professor Shigehiro Hirano, my "Guru" of chitin and chitosan chemistry who taught me all I had to know about the biopolymer; and Professor David F. Williams, who gave me my start in biomaterials.

viii

To my colleagues who have participated with me in my research over the past decade, Lee Yong Lim, Aileen Wee, Teck Koon Tan, Sek Man Wong, Garth W. Hastings, Willem Stevens, Suwalee Chandrkrachang, Kensuke Sakurai. To my students and staff, who toiled under their demanding taskmaster, Ling Fong Tay, Hwee Chze Li, Weng Keong Loke, Sandy Khoh, Juming Wong, Andrew Wan, Su Ching Tan, Wai San Loon, Weng Fun Ng, Christina Tan, Chue Feng Tee, Nealda Leila B.M. Yusof, Kok Sum Chow, Theingi Maw, Sarah Teng, Anne-Sophie Baguenard, Karen Teo, Tzehan Lim, Judy Saw, Lishan Wang, Zigang Ge, Marie Chan, Siok Lam Lim, Siok Ghee Ler, Fabien Cabirol.

To my colleagues and friends of the Chitin/Chitosan community who have shared their knowledge of our field so readily and freely. A special thanks to Professor RH Chen for providing an advance soft copy of his talk at the 8th ICCC prior to the release of the proceedings.

I would also like to express my gratitude to the agencies that have supported my research in funding and fellowship awards: The National University of Singapore, The British Council, The Association of Commonwealth Universities, The Japan Society for the Promotion of Science, The National Science and Technology Board, Singapore, The European Union, and The French Embassy, Singapore.

My sincere thanks to Nealda for acting as my editorial assistant, questioning, correcting and formatting the manuscript; Marie for acting as research assistant finding all the patents and typing them up; Lee Yong for reading the drafts providing her incisive comments and Khoon Seng for his industry perspective on Chapters 8 and 9. My deep appreciation to my Publishing Editor, Lucy Dickinson, who has been so wonderful and enthusiastic in making all the arrangements and accommodating all my requests in the course of publishing this book. To my wife, Val, and our dogs, Sprite and Tammy-Sue, for their patience while I was absent from our daily walks.

Finally, I thank GOD for placing the right events and people at the appropriate time in my life despite myself.

Eugene Khor
(Ph.D. Virginia Tech)
Singapore, May 2001

TABLE OF CONTENTS

CHAPTER 1: THE RELEVANCE OF CHITIN

1.1 CHITIN: FROM THEN TO NOW

Biopolymer is a term commonly used to refer to polymers biologically synthesized by nature. Polysaccharides are one such class of biopolymers, comprising of simple monosaccharide (sugar) molecules connected by ether type linkages to give high molecular weight polymers. Compared to their renowned cousins, the nucleic acids and proteins, the polysaccharides have traditionally been considered of less significance and regarded primarily as structural materials and metabolic energy sources. Only in the latter half of the 20th Century has interest in the chemistry of the polysaccharides witnessed a renaissance brought about by the development of new methods of isolation, extraction, separation, chemical and enzymatic modification coupled with the advent of sensitive and powerful instrumental analysis techniques. This has led to the "re-discovery", identification and functional elucidation of traditional and newly discovered polysaccharides, illuminating their inherent and essential roles in biological function.[1]

Among the polysaccharides, cellulose and chitin are the two most abundant biopolymers in the biosphere. Despite being scientifically discovered earlier than cellulose, chitin initially received limited attention. By contrast, extensive research and development has centered on cellulose since the late 19th Century.[2]

The first explicit account of chitin was in 1811, attributed to the Frenchman, Braconnot, Professor of Natural History, Director of the Botanical Garden and member of the Academy in Sciences of Nancy, France.[3] Braconnot described and named the alkali-resistant fraction from isolates of some higher fungi "Fungine". In 1823, Odier found a material with the same general properties as fungine in the cuticle of beetles and designated it "chitin" after the Greek word "chiton" that denotes "coat of mail" in reference to the cuticle. Subsequently, the chemical character of chitin began to be elucidated. The presence of nitrogen in chitin is attributed to Payen in 1843 while Ledderhose identified glucosamine and acetic acid as the hydrolysis products of chitin in 1876. Tiemann proposed that the amino sugar was based on either glucose or mannose and Fischer and Leuchs confirmed the attachment of the amino group at the C-2 position of the monosaccharide unit. Karrer and co-workers and Zechmeister and co-workers separately performed experiments to determine that N-acetylglucosamine was the primary constituent of chitin. Last, the (1-4) linkage and its β-D conformation was confirmed by X-ray diffraction, enzymatic and deamination reactions by Meyer and others in the first half of the 20th Century.

The discovery of chitosan is ascribed to Rouget in 1859 when he found that boiling chitin in potassium hydroxide rendered chitin soluble in organic acids. In 1894 Hoppe-Seyler named this material chitosan. Only in 1950 was the structure of chitosan finally resolved.

It has taken more than a hundred years to arrive at the chemical identity of chitin and the revelation of its polymeric properties. This is not surprising considering that the term polymer was not even acceptable until the 1930's onwards! Nevertheless, the necessary foundation was laid for the eventual emergence of chitin as an important biomedical material.

1

1.2 DEFINITIONS

Before proceeding further, it is necessary to define what chitin and its principal derivative, chitosan, is. The earlier literature is littered with examples of the casual interchange between these two terms and there have been instances when chitin is referred, where it is clear from a reading of the content, that chitosan was being discussed. Only in the past ten years, has there been more consistency in the usage of the terms chitin and chitosan. The definition and description that follows incorporates what is widely accepted by the chitin/chitosan community today.

The ideal structure of chitin is a linear polysaccharide of β-(1-4)-2-acetamido-2-deoxy-D-glucopyranose where all residues are comprised entirely of N-acetyl-glucosamine residues or in chitin jargon, fully acetylated. Likewise, the ideal structure of chitosan, the principal derivative of chitin is a linear polymer of β-(1-4)-2-amino-2-deoxy- D - glucopyranose where all residues are comprised entirely of N-glucosamine residues or fully deacetylated. The short-form notations for ideal chitin and chitosan are depicted in Figure 1.1 and employed throughout this book.

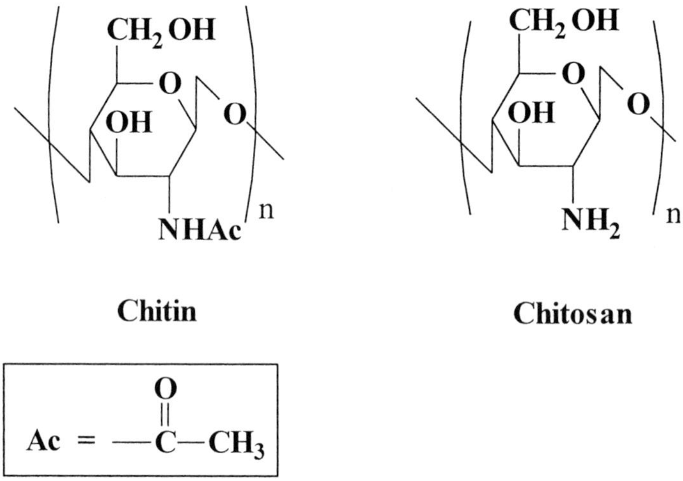

Figure 1.1:   Idealized representation of chitin and chitosan

Traditional sources of the biopolymer, however, do not result in 100% acetylated chitin or 100% deacetylated chitosan and in reality, the biopolymer exists as a co-polymer as represented in Figure 1.2, where the numbers in square brackets on the extreme left ring assigns the six carbons in the glucopyranose ring from C-1 to C-6. Specifically, the substitution at the C-2 carbon of the glucopyranose ring can be either with the acetamido group or an amino group. The differentiation between chitin and chitosan is to consider their respective acetyl content. When the number of acetamido groups is more than 50% (more commonly 70-90%) the biopolymer is termed chitin. In chitin terminology, the number of acetamido groups is termed the degree of acetylation (DA). On the other hand, when the degree of deacetylation (DD) or the amino group is predominant, the biopolymer is termed chitosan.

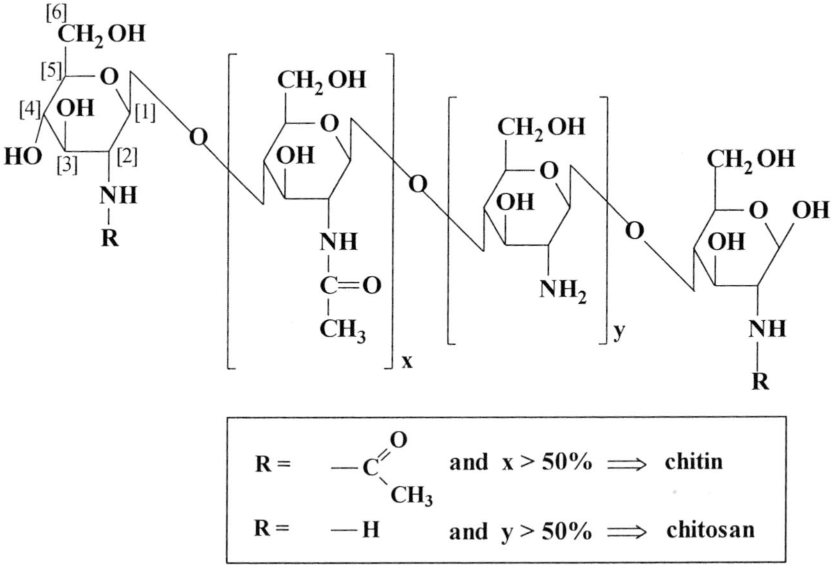

Figure 1.2: Chemical structural representation of chitin and chitosan depicting the co-polymer character of the biopolymers

Instant differentiation between chitin and chitosan can be attained from their solubility and nitrogen content.[4] Chitin is soluble in 5% Lithium chloride/N, N-Dimethylacetamide solvent [LiCl/DMAc] and insoluble in aqueous acetic acid while the converse is true of chitosan. The nitrogen content in purified samples is less than 7% for chitin and more than 7% for chitosan.

In this book, the term chitin will be used generally to refer to chitin and its derivatives, including chitosan. When direct references are made to chitosan or any chitin derivative, the specific names will be used. Chitin also has three crystalline forms, α, β and γ. Unless otherwise specified, chitin denotes α-chitin.

1.3 OBSCURITY

Despite its early discovery, and being the second most abundant polysaccharide after cellulose with an estimated annual production of at least 10^{10} tons per year in the biosphere, chitin has remained an almost unused biomass resource late into the 20th Century.[5] Why did chitin languish in obscurity for so long?

First, if we contrast chitin to cellulose, one explanation could be the source of the raw material. Cellulose is readily available from trees and cotton plants; two traditional sources that date back centuries, almost to the beginning of civilization. Coupled with the centuries of experience in exploiting and using this resource, it is readily understandable that cellulose has been well entrenched as a biopolymer of significant utility. Conversely, sources of chitin comprising of shells of crustaceans, insects and fungi, are more difficult to exploit and therefore, perceived to be of insignificant usefulness except as food and decorative ornaments. Second, looking back at the brief historical account of chitin and chitosan described above, the biopolymer belonged primarily in the realm of the naturalists, whose preoccupation lay in identifying and characterizing chitin and chitosan in plants and animals. These included isolating and determining the composition, the structure, macrostructure and physical properties and establishing their biochemical and biological roles.[6] A modicum of chitin chemistry only began to appear in the late 1930's. Very little else was reported in the ensuing years from its discovery until the 1950's. A third factor is the difference in the physical properties of chitin in comparison to cellulose where the strong hydrogen-bonding network in the former led to a momentary dead-end to its manipulation and use. The search for an effective solution to this situation was the beginning of a series of on-going efforts to defy the chemical and structural inertness of chitin.

The turnaround from anonymity can probably be attributed to two unrelated yet vital events. This was the advent of the canning industry that came into being in the 1800's. Canning enabled the preservation of food for extended periods of time. Obviously, the seafood processing industry of which crabmeat and other shellfish products were a part of embraced canning and grew. Coupled with an increasing consumption by a growing world population and the globalization of food taste, the shellfish waste that was generated gradually increased, eventually leading to the need to resolve this accumulating industrial waste problem in the 20th Century.

1.4 FROM WASTE TO NICHE MATERIALS

Commencing in the 1950's, a more concerted effort began to appear in the scientific literature. A natural course in finding a solution to the abundant shellfish waste from seafood processing is the interaction between industry and Academia. One example is Tottori, Japan, famous for its snowcrabs and associated seafood processing facilities that generated such shellfish wastes. It is probably no coincidence that Professor Shigehiro Hirano of Tottori University (now retired) became an eminent pioneer in chitin research since the late 1960's. This example duplicated in similar situations around the world from the 1950's onward no doubt has changed the course, or at least fast-forwarded the chitin story outcome. Eventually, this resulted in the increased understanding of the science and technology of chitin that culminated in the first "milestone" as it were, with Professor Muzzarelli's landmark book.[7]

Next began the giant leap in the study and application of chitin and chitosan arising from two fortuitous events. First, industrial interest turned into activity that established the regular availability of the biopolymer for research and commercial exploitation. Today, producers of chitin and chitosan span the globe and are found in North America, Europe and Japan and increasingly from China, India and South East Asia.[8] The quality of the product is varied. The predominant chitin and chitosan products are produced and consumed mainly in Japan. Increasingly, Korea is producing a mix of products aimed at an export market.

Second, the momentum generated by the series of International Chitin/Chitosan Conferences beginning with the first at MIT, USA in 1977 and with the 8th recently held in Yamaguchi, Japan in 2000. The European Chitin Society and the Japanese Chitin and Chitosan Society also further stimulated activity with their own meetings and the latest, The Asia-Pacific Chitin/Chitosan Symposium series, begun in the mid-1990s, adds to the on-going international forum on the biopolymer.

Application area	Specific use
Water treatment	Coagulating/flocculating agents for polluted waste waters
	Removal/recovery of metal ions from aqueous waste water
Agriculture	Plant elicitor
	Antimicrobial agents
	Plant seed coating
	Fertilizer
Textile and paper	Fibers for textile and woven fabrics
	Paper and film
Biotechnology	Chromatography packing
	Enzyme immobilizing materials
Food/Health Supplements	Natural thickeners
	Food additives including pet food
	Food processing (e.g. in sugar refining)
	Filtration and clarification
	Hypocholesterolemic agent (slimming agents)
Cosmetic	Ingredients for hair and skin care (conditioners)
Biomedical	Burns and wounds dressings for humans and animals
	Biomaterial (e.g. absorbable sutures)
	Anticoagulant or antithrombogenic materials (as sulfated-chitin derivatives)
	Hemostatic agents (as chitosan)
	Drug delivery, gene delivery

Table 1.1:   Applications of chitin, chitosan and their derivatives[9]

6

The driving force for much of the excitement surrounding chitin and chitosan are the potential applications that the materials can be used for. Table 1.1 lists some examples of the known and potential applications for chitin, chitosan and their derivatives that have caught the imagination of scientists, raw material producers and manufacturers and users alike.

Wastewater treatment using chitosan to chelate metal ions was one of the first applications of chitin. Various formulations for hair care products such as shampoos and conditioners followed suit. Over the past 20 years the other fields of applications have come on stream, key among them being the use of chitosan as hypocholesterolemic agents. Figure 1.3 shows a collage of some of these products.

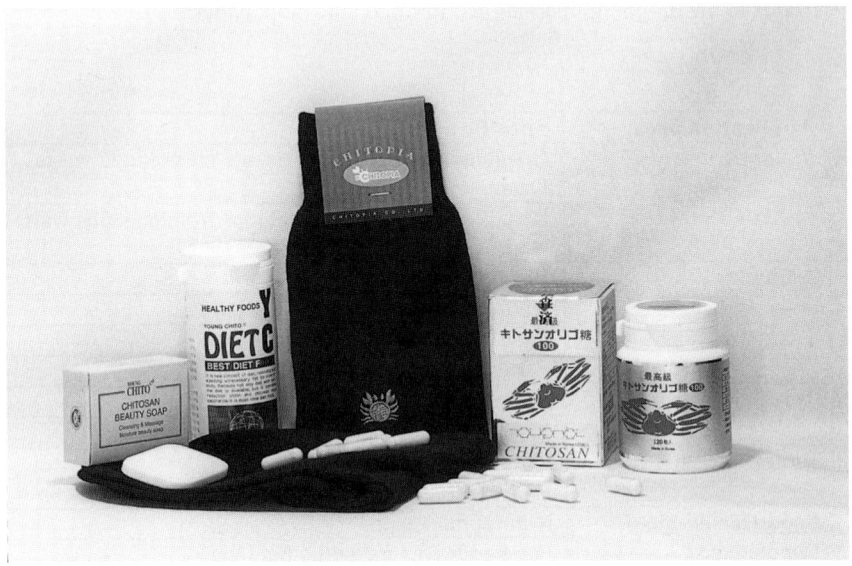

Figure 1.3:   Representative products of chitin and chitosan such as soaps, socks, weight-control pills and health food supplements [By permission of the Korea Chitosan Company, Ltd]

The market size estimate as listed in Table 1.2 illustrates the increasing potential and versatility of this biopolymer. It is noteworthy that biomedical applications have the potential to be the largest revenue generators.

Therefore, the flurry of activities commencing in the late 1970's reporting the "discovery" of this "new" and exciting biopolymer can be looked on as the watershed for chitin. From virtual anonymity, many products containing chitin and chitosan are now known and more are forthcoming. These activities have in turn lead to the need for a better understanding of the production, science and technology and utilization of chitin, sustaining the research cycle that is so necessary for the future applications and scientific knowledge of chitin.

Industry	Japan Market (10^6 US$/year)[10]	World Market (10^6 US$/year)[11]
Agriculture	-	2.3
Bio-fertilizer	21	-
Food and beverages	-	110
Cosmetics and toiletries	-	90
Biomedical	2000	1250
Immobilized cell and culture	-	45
Chitopack KQ 8025	3	-
Waste and drinking water treatment	-	140
Total	**2024**	**1915**

Table 1.2:   Potential size of the market for chitin, chitosan and their derivatives

Today the exciting promise of chitin has been delivered and more importantly, is set to play a significant role in the biomedical field as a biomaterial of the 21st Century. Perhaps chitin may even one-day surpass its older and better-known sibling, cellulose.

1.5 BIOMEDICAL APPLICATIONS

Towards the close of the 20th Century, the chitin community has become increasingly enthusiastic over the biomedical opportunities for this material. Why is this so? A plethora of processing methods including its chemistry and characterization together with a more in-depth knowledge of biomedical applications including cell/tissue interactions, have emerged. As a result, the breadth of biomedical applications spanned by chitin continues to expand, making chitin a material increasingly impossible to ignore. The possibilities appear endless. The surge of chitin in the biomedical direction is rooted from the postulation that there would be better acceptance of the material by the human biological system due to its natural origins and close analogy to body constituents. Together with the potential of limitless supply of this renewable material, these major influences have perpetually thrust chitin to the forefront as a biomaterial. Will this "gold-mine" of opportunity be realized?

The aim of this book is to address these biomedical applications of chitin. Each critical aspect that can shed light on these issues will be put forward. In the ensuing chapters, the foremost biomedical applications that have been proposed for chitin, wound dressings, blood-interaction materials, orthopedic applications, tissue engineering scaffolds and drug delivery will be surveyed. This will be followed by an assessment of whether chitin can and will meet the diverse biomedical applications proposed by carefully considering the requirements governing them. Surveying and placing in context all the relevant biocompatibility studies that has been reported for chitin so far and asks what else needs to be done.

For chitin to become a biomaterial of the 21st Century, the impact of the sources and production of chitin must also be carefully considered since they have impact on the biopolymer's availability for chitin to escalate in utility as a biomaterial. Equally important will be how the properties of chitin can be influenced by the mode of manipulation. The role of the many varied chemistries of chitin and their relevance in chitin exploitation will also be touched on. Finally a status report where chitin is on the regulatory road to approval will be discussed and a comparison with some of its competitors will conclude this exploration of chitin fulfilling its biomaterials role.

1.6 REFERENCES

[1] G.O. Aspinall, Polysaccharides. Pergamon Press, Oxford, UK, 1970

[2] Encyclopedia of polymer science and technology, Volume 3, H.F. Mark, N.G. Gaylord, N.M. Bikales, eds., John Wiley & Sons Inc., New York, N.Y., 1965

[3] C. Jeuniaux, A brief survey of the early contribution of European scientists to chitin knowledge. Advances in Chitin: 1, A. Domard, C. Jeuniaux, R. Muzzarelli, G. Roberts, eds., Jacques Andre Publishers, Lyon, France, 1996. 1-9

[4] R.A.A. Muzzarelli, Chitin. Pergamon Press Ltd, Oxford, UK, 1977. 87

[5] K. Kurita, Chitin and chitosan derivatives. in Desk Reference of Functional Polymers. Synthesis and Applications, R. Arshady, ed., America Chemical Society, Washington D.C., USA, 1996. 239-259

[6] M.Falk, D.G. Smith, J. McLachlan, A.G. McInnes, Studies on chitan ($\beta(1\rightarrow4)$-linked 2-acetamido-2-deoxy-D-glucan) fibers of the diatom *Thlassiosira fluviatilis hustedt*, II: Proton magnetic resonance, infrared and x-ray studies. Canadian J. Chemistry 44 (1966) 2269-2281

[7] R.A.A. Muzzarelli, Chitin, Pergamon Press Ltd, Oxford, UK., 1977.

[8] Chitin and chitosan: An expanding range of markets await exploitation, 3rd Edition. John Wiley and Sons Inc, New York, N.Y., 1998

[9] M.F.A. Goosen, S. Hirano. in Applications of chitin and chitosan, MFA Goosen, ed., Technomic Publishing Co., Inc., Lancaster, PA. 1997. Chapters 1 and 2.

[10] Liu 1994 Food Industry 26(1): 26-36 (in Chinese). Courtesy of RH Chen, Taiwan, 2000

[11] D. Knorr, Recovery and utilization of chitin and chitosan in food processing waste management. Food Technology 45 (1991) 114-122

CHAPTER 2: CHITIN BIOMEDICAL APPLICATIONS

2.1 BIOMEDICAL TECHNOLOGY

The alleviation of human pain and suffering is a noble endeavor. In the 20th century, the explosion in the advancement of science and technology has revolutionized the art of medicine hitherto unprecedented in the history of mankind. Today, sophisticated methods are available in treating an innumerable mix of human ailments caused by diseases and accidents, providing the patient with a quality of life that would have been impossible without them. One of the outcomes of this transformation has been the development of high-performance medical devices and the attendant use of materials in the treatment and/or replacement of damaged body tissue. In the process, the birth and growth of the biomedical technology industry that includes medical instrumentation, diagnostics and medical devices, has today burgeoned into a US$100 billion industry worldwide and growing at an annual rate of at least 10%. Most exciting of all, the biomedical technology arena is one where clinicians, engineers and scientists engage in a multi-disciplinary endeavor to continuously improve the offerings of medical technology for patient care.

It is pertinent at this juncture to identify what is meant by biomedical applications. Biomedical applications relate to the impact of a material, device, or procedure in a medical or clinical situation on the health care of humans. The expected outcome should be positive when properly utilized. The application can be a product as in medical devices such as a simple piece of gauze for cleaning wound and syringes, or as complex as pacemakers, orthopedic implants and artificial heart valves. Biomedical applications can also be a service for example the analyses of blood samples or the testing of a device for sterility. For the purpose of this book, we take a focused view by considering the biomedical applications that have been proposed for chitin. As will become evident, most of these biomedical applications will more or less be within the definition of a medical device except for drug delivery that normally falls under the purview of pharmaceuticals. Therefore, it is appropriate to define what a medical device is.

"A medical device is defined as an instrument, apparatus, implement, machine, contrivance, implant, *in vitro* agent, or other similar or related article, including a component, part, or accessory that is:

1. Recognized in the official National Formulary, or the *United States Pharmacopoeia* (USP), or any supplement to them.
2. Intended for use in the diagnosis of disease or other condition, or in the cure, mitigation, treatment, or prevention of disease, in man or other animals, or
3. Intended to affect the structure or any function of the body or man or other animals, and which does not achieve any of its primary intended purposes through chemical action within or on the body of man or other animals, and which is not dependent upon being metabolized for the achievement of any of its principal intended purposes (CDRH, 1972)."[1]

For a chitin biomedical application to be valid, chitin must be used as an integral component of the device or purpose of use. As medical devices, the applications of chitin can be conveniently divided into 2 classes, external and internal. Likely applications of chitin as external devices would be external communicating devices that come into contact with intact natural channels of the body such as the eye, vagina and the gastro-intestinal tract and those

9

that breach the body surfaces or contact blood such as in intravenous cathethers or conduits for fluid entry. Examples of chitin applied in external medical devices are contact lenses, wound dressings, hemostatic agents and coatings of the inner lumen of blood-contacting tubing.

Internal devices are normally implants that are targeted for bone, tissue and tissue fluid and blood. Examples of internal medical device applications for chitin include orthopedic implants such as bone pins, plates and cements, tissue engineering scaffolds, systemic anticoagulants, drug delivery components and gene delivery vehicles, the last two examples crossing the boundary into pharmaceutical applications.

A casual survey of chitin publications reveal a staggering volume of scientific reports and over 600 US patents since 1975, not including European patents and the ever-increasing number of Japanese patents. This is a testimony to chitin that so many diverse opportunities have been evaluated and proposed possible. The considerations of chitin for biomedical applications date as far back as the beginning of the emergence of chitin as a material to be reckoned with in the late 1950's. While it would be conceivable to document all of these (a monumental task in itself), it is the theme of this book to focus on specific areas based on the scientific relevance, utilization and probability of success, of the applications. Accordingly, the following survey will discuss wound dressings; blood interactions; orthopedic implants; tissue engineering and drug delivery.

2.2 WOUND DRESSINGS

Skin is an organ that covers about 2 square meters of the human body's surface. Some of its more important roles are regulating body temperature, providing a barrier to disease and removal of body waste.[2] A wound arises when skin is compromised by an injury as a result of mechanical trauma, surgical procedures, or from pressure sores and burns. The wound causes physical, mechanical and thermal damage of the skin surface that can lead to disruption in the physiological functions of internal tissues resulting in acute body dysfunction. The wound healing process is complex and encompasses a continuum of overlapping phases that includes hemostasis, inflammation, proliferation, granulation and remodeling.[3]

Wound dressings are used to protect the site of injury from further insult, contamination and infection that may impede healing. The ideal wound dressing would also facilitate and accelerate wound healing. Today, there are many wound dressings available on the market that addresses different kinds of wounds, treatments and phases in the wound healing process. Dressings are fabricated from both synthetic and natural materials, and now include tissue-engineered skin substitutes. With an increasing understanding of the science of wound healing and wound repair at the molecular level, innovations will continually be introduced in the years ahead.

2.2.1 Chitin-Based Wound Dressings

The wound dressing application is by far the most comprehensively evaluated biomedical application for chitin, touted as one of many natural materials with wound healing augmentation properties. What is chitin endowed with for it to be advocated as a good wound care material? The origins for chitin being propounded as a candidate for wound healing can be traced back to the breakthrough paper by Prudden et al.[4] Based on their study of the use of cartilage in accelerating wound healing, they deduced that the active component was N-acetyl-glucosamine. To verify their hunch, chitin obtained from shrimp and fungal sources were applied as topical powders on wounds. Eventually, results confirmed chitin's accelerating effect in wound healing. The authors proposed that the chitin powders released N-acetyl-glucosamine as a consequence of the breakdown of chitin by the enzyme lysozyme, abundantly present in fresh and healing wounds. It is of significance to note that fungal chitin was resorbed twice as fast as shrimp chitin.

Progressing from the sprinkling of chitin powder, more formal methods expectedly ensued. In the form of chitosan films, favorable attributes for wound healing include the ability to form tough, water-absorbent and biocompatible films with good oxygen permeability.[5] Using a rat wound model, Allan et al showed that survival of animals was improved when the wounds were treated with chitosan films in various configurations of molecular weights and in some samples, with the inclusion of a silver antibiotic.

Equally comparable results were obtained in the study that utilized chitin as a non-woven fabric-type dressing. The non-woven dressing was prepared by first making chitin fiber, cutting the fiber to desired length, dispersing the cut fibers in water and binders giving non-woven sheets that was cut to dimensions suitable for a dressing. These non-woven chitin dressings were shown to be effective in treating burns and skin ulcers, skin-graft areas and dressing of donor sites, in some instances accelerating epithelialization and granulation in a sampling of 91 human subjects.[6] In addition, the wounds were kept dry and the dressing adhered to the wound well. The direct use of fungal mycelia to produce a wound dressing has also been attempted.[7] A non-woven mat was obtained by first processing the mycelia to remove protein and pigment, followed by isolation of fibers in the 10-50 μm diameter range and final consolidation into a freeze-dried membrane under aseptic conditions. The wound healing of this fungal-based non-woven mat as surmised from wound contraction measurements on rat model studies were favorable.

Chitin can also be prepared in the water-soluble form (WSC) by carefully deacetylating to about 50% N-acetyl content.[8] In a comparative study of chitin, chitosan and WSC powders and WSC solution, wounded skin treated with WSC solution was found to have the highest tensile strength. The healing rate was also fastest for WSC solution followed by WSC powder, chitin powder and last, chitosan powder. Fluid absorbing chitin beads has also been proposed as a wound dressing material.[9] Chitin beads were first obtained by coagulation in ethanol followed by a carboxymethylation step to give the beads reversible fluid absorbing properties useful for the absorption of wound exudates. Finally, the preparation of a bi-layered chitosan membrane by "immersion-precipitation phase inversion" has appeared.[10] This membrane has a thin layer of chitosan that acts as an antibacterial and moisture control barrier attached to a sponge layer that can absorb wound exudates. The membrane adhered well to the wound surface and promoted wound healing normally observed with chitosan-based dressings.

Departing from just utilizing chitin and chitosan, Muzzarelli et al became proponents for a chitosan derivative, N-carboxybutyl-chitosan that they developed. The advantages listed for N-carboxybutyl-chitosan were its water-solubility making processing easier, its gel forming ability permitting the absorption of wound exudate and its ease of sterilization. In rabbit animal model studies, several favorable factors for wound healing were found including the formation of repair tissue and the absence of scar formation and contraction.[11] Similar results were obtained when the study was extended to human patients.[12] Last, in comparison to fibrin glue, the superior performance of N-carboxybutyl-chitosan was again manifested.[13] The healing rates of both materials were similar, but in the restoration of biological activity, N-carboxybutyl-chitosan was found to direct a more regular histoarchitectural reconstruction of the tissue.

The combinatorial effect is another channel that must be explored to exploit the benefits of individual materials while minimizing their limitations and is no surprise that this would occur with wound dressings. Chitosan has been used in combination with collagen and glycosaminoglycans (GAG) in a clinical setting.[14] Inspection of the wound bed after 10 days disclosed controlled vascularization as a prelude to epithelialization with histology indicating organized dermal repair. While this material did not match the superior wound healing qualities of an autograft, the authors proposed that the dressing should provide sufficient healing in situations where auto- or homo- grafts were not available. Heparin, another glycosaminoglycan has also been combined with chitosan to form wound-healing membranes, including a gel-form incorporating methylcellulose.[15] Using an *in vitro* model based on human skin, the gel membrane healed the wound best, followed by the chitosan-heparin membrane and the poorest healing obtained with methylcellulose.

The inclusion of antimicrobial agents into wound dressings is another strategy that has been investigated. In one preparation, β-chitin was combined with polyethylene glycol to form a partial gel.[16] Silver sulfadiazine was next added to the partial gel, with subsequent precipitation in a non-solvent producing the combination gel that was finally freeze-dried to give the dressing. Results from animal studies indicated infection controlled wound healing. In another study, a chlorhexidine containing chitosan-based wound dressing was shown to have antibacterial efficacy towards the primary wound bed bacteria, *Pseudomonas aeruginosa* and *Staphyloccocus aureus*.[17] The bilayered dressing was fabricated by combining two separate films, a carboxymethylated chitin hydrogel that provided the exudate-absorbing component pressed together with a chlorhexidine loaded chitosan film to give the dressing.

Apart from demonstrating that chitin and chitosan are good wound healing materials, their performance in comparison to other wound dressing materials has also been evaluated.[18] In one *in vitro* study, four chitosan-based materials and three commercial samples were evaluated with Swiss 3T3 fibroblasts. Methylpyrrolidinone-chitosan was found to be the best chitin-based material (Table 2.1) comparable to the commercial non-woven calcium alginate fiber.

Wound healing evaluation of chitin and chitosan granules referenced to untreated controls has also been performed.[19] Macroscopic inspection of the wound site showed complete re-epithelialization at 28 days post procedure in 100% of sites treated with chitin, 75% of sites treated with chitosan and less than 50% for the controls. However, when analyzed statistically, the differences in these macroscopic observations were insignificant suggesting the subjectiveness of macroscopic scoring evaluations. On the other hand, the histology assessment based on the presence of inflammatory cells, fibroblasts and neovasculature detected at the wound site was very low for chitin and chitosan compared to controls, indicating normal healing occurring.

As interest in the use of chitosan for wound dressings increased, more comprehensive justification based on chitosan's ability to promote wound healing was sought. Using a bovine animal model, Minami et al implanted commercially available cotton-type chitosan

Material	Rank*
Methylpyrrolidinone-chitosan	3
Chitosan glutamate	11
Chitosan lactate	7
Chitosan chloride	12
Collagen fleece	6
Non-woven calcium alginate fiber	3
Gelatin sponge	7

Table 2.1: Cell culture compatibility ranking of wound dressing materials[20]
* Based on sum of data obtained from cell growth, logarithmic growth phase and cell confluency; a lower number is translated as more compatible

into subcutaneous tissue. Histological results showed the presence of polymorphonuclear (PMN) cells, necessary for phagocytosis of the initial unorganized collagen fibrils, accumulated at the chitosan fiber regions after 7 days.[21] At day 14, connective tissue reconstruction accompanied by the complete disappearance of the PMN cells was observed. The authors concluded that chitosan stimulated the migration of PMN cells to a wound site to accelerate wound healing. The presence of matrix metalloproteinases (MMP) during wound healing is also important as MMPs play a vital role in the digestion of protein tissue initially formed in the remodeling, facilitating regeneration and remodeling of the wound.[22] Nakade and co-workers showed the level of MMP for wounds treated with chitin sponge was much higher than control, suggesting chitin's ability to activate vigorously the cellular response necessary for good wound healing.

2.2.2 Chitin-Based Wound Dressings Patents

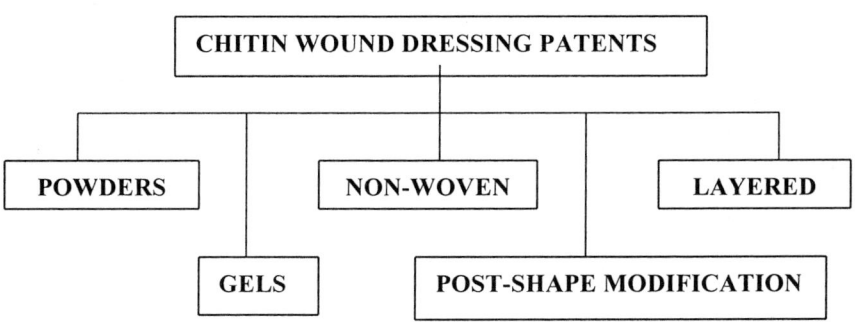

Figure 2.1: Wound dressing patent development

In tandem with the development of the science of chitin-based wound dressings, patents accompanied most new methods devised. The patents to utilize chitin in wound dressings are varied but fit five broad strategies summarized in Figure 2.1. Naturally powders were the first to be patented in the 1970's as the knowledge to manipulate chitin and chitosan was just beginning to be explored.

Balassa, one of the co-authors of the original work by Prudden et al, was awarded several patents describing how chitin and its derivatives, applied primarily not only as a fine powder, but also incorporated into fibers, sutures, components of gauze etc. could be used in wound healing.[23] Malette and Quigley refined this initial work by defining the use of chitosan in their wound treatment invention that included solution, powders, films and mats.[24] The primary concern of the invention was to promote hemostasis and prevent fibroplasia thereby enhancing tissue regeneration. Although studies with a rat model found chitosan mats and solutions to be ineffective compared to the controls, good results were found with mongrel dogs where chitosan solutions used to treat wounds on the skin, liver, spleen and bone marrow promoted wound healing, with only thin scarring. Subsequently the limitations of applying only chitin were indicated.[25] In this newer invention, partially deacetylated chitin was prepared as spherical particles bonded with proteases such as chymotrypsin, trypsin and papain. The partially deacetylated chitin permitted swelling that was useful for absorption of exudates while the proteases decomposed unwanted proteins that were present. However, no study data accompanied the patent. Subsequently, inventions that used powders appear to fizzle out.

Non-woven fabric dressings is another popular method of producing suitable chitin forms as wound dressing as the technique is straight forward, such as that described in the production of short chitin fibers from chitin-dope solutions followed by combining with different binders including vinyl acetate and carboxymethylcellulose to give chitin-based non-woven wound dressings.[26] Many chitin derivatives were mentioned as substitutes for chitin in the patent but no data on the effectiveness of the dressings were presented. In a 1990 invention, the dispersion of chitin to form wound dressings was described where the mild shearing of chitin with aqueous hydrogen peroxide or bleaches followed by treatment with sodium hydroxide, washed and finally dispersed in water gave a paper like dressing upon drying.[27] The direct use of chitin obtained from fungal mycelia that was grown, harvested, treated to remove protein and cellular material, and finally processed directly into freeze-dried forms such as non-woven mats, sheets and rolls was the subject of an invention by Sagar et al.[28] A remark in the patent indicated that encouraging healing results with the dressing were obtained although the biological model was not identified. Chitosan was also noted as a component of a gel-based wound dressing in combination with gelatin films suitable as dressings when the two components were blended from weight ratios of 1:3 to 3:1 with a plasticizer, processed and dried. Pig model studies indicated the efficacy of this wound dressing in limiting wound contraction for all ratios.[29]

With progress, the design of exudate absorbing dressings became the next trend. In this category, chitosan-glycerol-water gels as a vehicle to carry medicaments for wounds can be assigned as one of the forerunners.[30] Gel-like pastes comprising chitosan blended with hydrocolloid materials such as polysaccharide gums has also been described as wound-filling compositions.[31] When the chitosan content in the paste was increased, an antimicrobial effect was observed. The dispersion of chitosan and water absorbent polymer in latex foam can be seen as another step in the progression of this path of development.[32] In this invention, latex materials that can be foamed were blended with insoluble, but water absorbent polymers and chitin to give a wound dressing that had exudate absorbing and wound healing characteristics. In a related approach, polyvinylpyrrolidone was blended with chitosan or its derivatives in suitable ratios to give highly water swellable gels that were applied as a wound dressing.[33] A further variant to the gel wound dressing repertoire was a hydrogel product based on a three-component gel forming system.[34] This system comprising of a water-soluble polymer that imparts adhesion to the skin surface and a crosslinking polymer to hold the gel and chitosan or its derivative, was indicated for deep wound cavities.

Foam-sheet wound dressings with a water-soluble polysaccharide as its main component has also been patented.[35] Using chitosan in one variation, the foam was mechanically generated creating gas bubbles in the solution that upon drying gave a sheet that had sizable and widespread pores throughout the sheet. The presence of chitosan was as a wound-healing agent, while the water absorbent polymer facilitated exudate absorption. In a further variation, the inclusion of chitosan to exploit its bacteriostatic and hemostatic properties was also mentioned.[36] The primary component in this invention was alginic acid into which chitosan was blended to give several forms such as multi-layered composites, fibers and gels. Finally, a water insoluble but swellable wound dressing made from propylene glycol alginate and using chitosan as a crosslinking agent rounds off this category of chitin-based wound dressing inventions.[37]

The preparation of chitin into a fixed shape with subsequent chemical treatment appeared as a patent in 1987. In the description, chitin was first prepared as powders, fibers and non-woven fabric followed by treatment with alkali to give a chitin material that swelled about 10 times while retaining its original shape.[38] Khor et al later termed this process post-shape modification when they prepared chitin films, followed with surface carboxymethylated of the chitin film to give a water swellable reversible gel.[39]

The introduction of multi-layered wound dressings was inevitable, conceived to be bi- or multi functional with at least one layer for wound contact and another acting as the protective and/or moisture controller layer. In an early description, chitin was combined with extended poly-tetrafluoroethylene (ePTFE) in a bi-layer dressing. Alternatively, the chitin was prepared as a solution that penetrated the porous ePTFE layer to give a composite dressing.[40] A later version described a wound covering material that utilized a moisture-regulating polymer in contact with a wound covering gel-forming material among which are, chitin and chitosan derivatives, in the form of non-woven cloths, knitted cloths or porous membranes.[41] It was noted that the chitin-based samples brought about moderate healing in a rabbit wound model. In another multi-layered wound dressing patents chitin (more appropriately chitosan) was utilized for its blood clotting property and was applied as powder on the wound contact surface of the wound dressing with a water-soluble binder.[42]

In the 1990's, the knowledge of wounds and features required in materials that promote wound healing became more established. Coupled with an increased understanding of chitin, more superior methods were devised such as the combination with other wound promoting materials or the use of chitin derivatives with better properties and ease of processing. A case in point was the use of N-acylchitosans in combination with collagen where the combined materials could be processed into gels, or freeze-dried to give sponges and films.[43] Favorable wound healing on human subjects was reported. Next came the introduction of methyl-pyrrolidinone chitosan by Muzzarelli describing the preparation of the substituted chitosan and subsequently, the wound dressing.[44] The methyl-pyrrolidinone chitosan based wound dressing was found to advance wound healing due to its high susceptibility to lysozyme action at the wound site. The use of β-chitin derived from squid, laminated to a fish derived collagen, formed a wound dressing that encouraged cell adherence and proliferation.[45] β-Chitin was preferred because of its ease of processing compared to α-chitin. A tri-combination material invention was also the focus of the patent that described the processing of collagen with chitosan followed by addition of chondroitin sulfate and made into an artificial skin covering.[46]

The interaction of chitosan's cationic properties with iodine vapor led to the development of charge transfer complexes with high iodine loading on the biopolymer.[47] This is useful as iodine is a good disinfectant. The solubility of the biopolymer complex in aqueous acids, facilitated the preparation of powders, solutions and ointments. A heterologous skin substitute based on a chitosan foil containing glycerin as an elasticizing agent with pores for gas exchange, has also been patented.[48] Chitosan with a high degree of deacetylation, mixed and ionically bound with heparin or heparan sulfate to produce beads films, ointment and wound powder can also be prepared for wound treatment.[49]

N, O-Carboxymethyl-chitosan has also been formulated as a post-surgical lavage to wash wound areas thereby minimizing wound adhesion.[50] The use of chitosan as a selective super antigen absorber has also been described. Urea and thiourea moieties that have good antigen absorbing properties are chemically bonded to the amine functionality of chitosan. This gives a material that can be used as a wound dressing in fabric and film forms. The choice of chitosan was to preserve the good antigen affinity property after sterilization.[51]

Today chitosan-based wound dressings are a reality as attested by the fibers and gauze forms depicted in Figure 2.2.

Figure 2.2:   Chitosan Fibers and Gauze

In summary, from the survey of the scientific studies and patents on chitin based wound dressings, early inventions capitalized on the fledgling scientific principles of chitin and chitosan in wound healing i.e. N-acetyl-glucosamine. The practical methods of handling chitin and chitosan at the time included turning them into various forms of powders and dispersions. When the hemostatic and bacteriostatic properties became known, chitin foam, gel and laminate wound dressings emerged. In moving from powders to films, the effect of N-acetyl-glucosamine on wound healing was reconciled to the requirements of a good wound covering material. Today, some of the underlying principles of why chitin-based materials promote wound healing have been elucidated leading to the use of derivatives as well as processing or combining with other materials possessing properties relevant to wound dressings and healing.

What can be also garnered is while chitin and chitosan alone or in combination with other agents can provide wound healing, the ideal chitin-based wound dressing is still wanting. It has been stated by Damour et al that the recovery is not as ideal as that achieved with a homograft. In this respect cell seeding of chitin materials i.e. chitin-based tissue engineered skin substitute may perhaps, provide the closest thing to a homograft this side of technology can offer. This strongly implies that chitin wound dressing development faces an open road as technological boundaries are pushed with each new invention. Not surprisingly, novel ideas are still perpetuating in this, one of the most established areas of chitin.

2.3 CHITIN-BLOOD INTERACTIONS

Anticoagulation therapy is an essential component of open-heart surgery and kidney dialysis. This requires an agent that prevents blood from clotting to be administered during these procedures. Heparin, a naturally occurring sulfated glycosaminoglycan obtained commercially from the porcine intestinal mucosa, is the standard anticoagulant used clinically. Heparin is comprised of 3 saccharide units, β-D-glucuronic acid, α-iduronic acid, and β-N-acetyl-D-glucosamine (Figure 2.3). The molecular weight of heparin, ratio of glucuronic to iduronic acids and the number of sulfate group play an important role in the anticoagulant activity.[52]

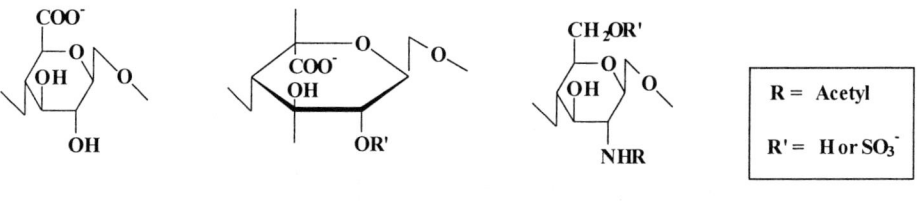

β-glucuronic acid   α – iduronic acid   N-acetyl-glucosamine

Figure 2.3   The key saccharide constituents of heparin β-D-glucuronic acid, α-iduronic acid, and N-acetyl-D-glucosamine

The similarity of chitin to heparin coupled with the work in the chemistry of derivatizing chitin, in this instance with sulfate groups, prompted much of the early interest in chitin as an anticoagulant. Quite fascinatingly, while chitin possesses the anticoagulant or hemocompatible property, it was later found that chitosan displays the opposite hemostatic or blood clotting property.

As far back as the early 1950's, Wolfrom et al reported the anticoagulant activity of chitin by heterogeneously sulfating N-deacetylated chitin (i.e. chitosan) with a chlorosulfonic acid/pyridine mixture. The *in vitro* anticoagulant activity of the sulfated chitosan was found to be 56 I.U./mg (international units/mg) with an accompanying higher toxicity compared to heparin, attributed to the high molecular weight of the sulfated-chitosan.[53] Subsequently, Wolfrom and Shen-Han sulfated chitosan at the N-2, O-3 and O-6 using two different methods, chlorosulfonic acid/dimethylformamide and SO$_3$/pyridine as sulfating agents. The resulting anticoagulation activity was around 50 I.U./mg regardless of synthesis method. Interestingly, the chlorosulfonic acid sulfating agent was found to give a total sulfate count of four on the saccharide ring, two on the nitrogen and two on the oxygens with a toxicity value of LD$_{50}$ 380, while the SO$_3$ sulfating agent gave a total sulfate count of two, one on nitrogen and one on oxygen with an LD$_{50}$ of 775 close to heparin.[54]

At about the same time, Warner and Coleman showed that exclusive sulfation at only the N-2 site obtained with a SO$_3$/pyridine mixture under alkaline conditions did not give any anticoagulant activity.[55] Whistler and Kosik later confirmed that O-sulfation was required for increased anticoagulant activity.[56] Finally, Horton and Just synthesized O-carboxylated, N-sulfated chitosan and reported the derivative's anticoagulant properties.[57] Muzzarelli et al also reported their sulfation studies on N-carboxymethylated-chitosan and its relationship to anticoagulant activity.[58] In a chitin-based study, Tokura et al prepared 6-O-sulfated-chitins with minimal N-sulfation that was found to delay thrombin activity, one of the indicators for good anticoagulation property.[59] Tokura et al also extended their work to the study of red blood cell binding.[60] By the mid 1980's however, dwindling interest in sulfating chitosans appeared to have set in, and acts as a good demarcation point for the first generation phase of investigating chitin as an anticoagulant.[61] Some possible explanations can be put forward, the first being the difficulty in handling the SO$_3$ sulfating agent, a rather nasty chemical system very often depolymerizing the biopolymer; the happy discovery that simple reactions generated acyl- and alkyl-chitosans derivatives found to be bestowed with good anticoagulant properties came into being and finally, the advent of a new interest of the blood clotting or hemostatic properties of chitosan.

One of the early studies in the change of direction, attributed to Malette and co-workers, was when they reported that whole blood formed a coagulum rapidly upon exposure to chitosan.[62] In subsequent experiments using a dog model, they found that the chitosan incorporated onto Dacron grafts formed coagulum, completely sealing the interstices of the graft. Chitosan of suitable deacetylation, soluble in distilled water and acetic acid was introduced intravenously to the site of bleeding to initiate hemostasis or selective tissue death.[63] Fradet et al confirmed these observations, demonstrating the hemostatic effect even in the presence of extensive anticoagulation therapy.[64]

The advent of non-sulfate chemical derivatives of chitin preparation methods were established in the mid 1970's and it was inevitable that studies to determine the presence of hemocompatible properties of these chitin derivatives would also start appearing. Kaifu and Komai reported the anticoagulant properties of a series of acylated-chitins derived by reacting chitin dissolved in methanesulfonic acid with the respective carboxylic anhydride.[65] All substituted chitins including diacetyl-chitin delayed the on-set of coagulation 2-3 times longer than chitin. Hirano and Noishiki subsequently prepared N-hexanoyl-chitosan and found it to be very hemocompatible while at the same time re-confirming the hemostatic property of chitosan.[66]

The suggested cause of blood coagulation with chitosan was the possible formation of polyelectrolyte complexes (PEC) between the amino functionality on chitosan and cellular elements of blood that contained acidic groups. This blood clotting inducing capability of the amino functionality is reduced upon chemical derivatization since the number of free amino groups available is diminished automatically that is further augmented by steric hindrance provided by the new substituent. This could well be the situation as the study of the clot-inhibiting profile of PEC formed by reacting oppositely charged polymers of chitosan and dextran sulfate (DS) also suggests.[67] When dextran with a molecular weight of 6000 was combined with chitosan clot formation was delayed when the order of forming the PEC was DS into chitosan. The reverse order of addition, chitosan into DS resulted in blood clotting quicker. It was concluded that the more abundant DS obtained in the DS into chitosan PEC formation process exposed a smaller profile of chitosan to blood, sequestering the interaction of chitosan with potential coagulating components in blood, therefore the better anticoagulant behavior of chitosan under these circumstances. When chitosan was permitted to display its usual character as in the reverse order of chitosan into DS PEC formation, blood coagulation occurred.

More recently, Wan et al revisited the N-acyl-chitosan derivatives blood compatibility issue, restricting the degree of acylation in the 20-50% range to avoid gel formation of the chitosan derivatives.[68] N-hexanoyl-chitosan was again found to give the best anticoagulant activity but in this instance, also found to be degradable by lysozyme, a common enzyme used in degradation studies of chitin.

Dutkiewicz et al added some spice into the action when they reported in the late 1980's the surprising anticoagulant properties of chitosan.[69] Their findings, based on whole blood clotting time (WBCT) and clotting time ratio (CTR referenced to glass), showed that chitosan unexpectedly exhibited a longer time to clot than acetyl-chitosan (regenerated chitin) and dibutyryl-chitin. To further complicate matters, their subsequent studies using an animal model showing chitosan playing a hemostatic role, contradicted their own earlier results. This apparent reversal was rationalized by the possible deactivation of factors XII and XI by chitosan in the intrinsic pathway of the clotting cascade.[70] It is important to note therefore, that the anticoagulant activity of chitosan has so far been achieved only *in vitro*. Most recently, Hoeskstra et al have demonstrated that microcrystalline chitin can also be used as an effective hemostatic agent.[71]

Interest in the blood interaction applications of chitin and chitosan appears to have waned in the 1990's as other biomedical applications in particular, drug delivery took off. This is not surprising considering that blood is possibly the most challenging of all human tissue and organs to deal with. A perusal of the patents is a further indication. Most patents describing chitosan in blood contact applications utilize chitosan as a coating on the surfaces of polymers onto which an antithrombotic agent is deposited.[72] In one patent description, the surface of the polymer was first primed by plasma polymerization onto which chitosan was coated. The chitosan coat was exposed to a heparin solution wash, presumably promoting deposition of the heparin onto it. A final wash of this surface gave the antithrombotic coating. Similarly, the formation of a thromboresistant surface, utilizing chitosan as a base membrane, combining with polyvinyl-alcohol or other blood compatible material was described in a recent coating invention.[73]

Amiji has studied the blood-compatibility of chitosan-polyethylene (PEO) blends as membranes for hemodialysis, where the chitosan-PEO was found to be effective in providing the hemocompatibility required.[74] The choice of chitosan was apparently due to its possible role in decreasing complement activation by the alternative pathway, attributed to the amino group. It is further noted that chitosan promoted platelet adhesion, an indication of the thrombogenic capacity. Therefore chitosan does appear to possess both anticoagulant and hemostatic effects and the issue is one of moderating these effects. When chitosan alone is utilized, in contact with whole blood it acts as a hemostatic agent i.e. blood will clot as chitosan activates platelet adhesion. When chitosan is combined with a blood compatible material such as dextran or PEO as discussed above, this effect is masked and blood compatibility is obtained. In this instance, the possibility of chitosan deactivating factors XII and XI may occur but as has been pointed out, this has only been observed *in vitro* and can only be resolved with further work. Finally, in another variation, chitosan was impregnated with anionic polysaccharides such as dextran sulfate to achieve blood compatibility and appears to hold this conclusion out.[75] The description of making chitosan biocompatible in a chitosan based blood compatible system for hemodialysis has also been patented.[76]

It is evident from the foregoing description that the blood interaction potential of chitin and chitosan enjoyed a period of promise that has apparently dissipated. If this is an accurate assessment, does this spell the end for chitin as a hemocompatible material and chitosan as a hemostatic agent? That chitosan is an established hemostatic agent is without question, so is the hemocompatiblity of sulfated-chitin derivatives. How can these properties be exploited from hereon?

For chitosan, the route is straightforward, embarking on a commercialization program to establish chitosan as a hemostatic agent. For the anticoagulant challenge, the first order of business is to reliably synthesize the required sulfated-chitin derivatives. The practice of chitin chemistry has been refined over the years (as will be elaborated in Chapter 7) and the synthesis of sulfated-chitin derivatives should be revisited with these new methods in the undertaking to re-establish chitin as a blood-compatible material.

2.4 BONE SUBSTITUTES

Bone is largely made up of two components, an intimate combination of collagen and calcium hydroxyapatite.[77] Many clinical situations require replacement materials to fill bone defects.[78] In any approach to address this issue of deriving bone substitutes utilizing synthetic materials, simulating the basic components of bone is a logical starting point. Calcium hydroxyapatite is a well-investigated ceramic material and there are many commercial sources for calcium hydroxyapatite meeting biomedical requirements. The role of collagen in bone is much like a soft pliable matrix onto which the hard hydroxyapatite is deposited. There are many suitable candidate materials that can take the place of bone collagen including extracted collagen from animal sources, alginates, polyhydroxybutyrate and of course chitin.[79]

2.4.1 Chitin-Based Bone Substitutes

Chitin has been applied both neat as well as in combination with calcium compounds in orthopedic applications. Maeda et al. were one of the first to use chitin in the form of braided filaments, rods and powders. These substitutes were found to be potentially suitable for sutures and temporary artificial ligaments for the knee joint.[80] As time progressed, the possible osteogenic, osteoconducting and osteoinducting properties became the subject of more thorough investigation. Borah et al studied the bone induction properties of N-acetyl-chitosan, finding that calcification at sites containing chitosan was constantly better than the control and concluded that chitosan had osteogenic properties.[81]

In several separate reports, Muzzarelli et al studied the utility of chitosan and its derivatives for orthopedic applications. In one study, N-carboxybutyl-chitosan was injected as a 2% solution into the meniscus region of the rabbit knee. After 45 days, the meniscal tissue site was found to exhibit structural repair processes.[82] In another study, imidazole-chitosan, a substituent that stimulates tissue reconstruction, was implanted into holes drilled in the femoral condyle site of sheep.[83] Histological findings indicated bone formation and that the chitosan material was osteoinductive as it promoted mineralization. Muzzarelli et al have also looked at the osteoinducting properties of hydroxyapatite nails surface coated with chitosan.[84] In this work using a rabbit model, chitosan adequately acted as a go-between for hydroxyapatite and bone and chitosan was declared osteoconductive. Bone morphogenetic proteins (BMP) have also been combined with N, N-dicarboxymethyl-chitosan by a solution based polyelectrolytic complexation (PEC).[85] Histological results based on implant studies in the femoral condyle of rats indicated that bone differentiation was more evident in explants that had contained the PEC. Recently, 6-oxychitin has also been evaluated for its bone regeneration properties using N, N-dicarboxymethyl-chitosan for comparison.[86] 6-Oxychitin seeded with osteoblasts showed the better osteo-architectural reconstruction compared to all other combinations used, despite an accompanying slower rate of healing.

Kawakami et al investigated the potential of a chitosan-hydroxyapatite paste as candidate bone substitute materials on a rabbit model.[87] One of their more interesting results was the finding at the implant site tissue, the presence of capillary rich connective tissue throughout the study period, indicating osteoconducting properties. However, this osteoconducting property could not be attributed singly to either chitosan or hydroxyapatite but only to the combination. Subsequently, Kafrawy et al looked at mixtures of chitosan sol, made with either malic acid or malonic acid, combined with β-tricalcium phosphate, ashed bovine bone and calcium hydroxyapatite.[88] The acids present in the sol decomposed β-tricalcium phosphate but not bovine bone or calcium hydroxyapatite and therefore, verifying the suitability of chitosan-based filler pastes. Wan et al have prepared calcium-containing chitin composites by inducing the precipitation of calcium phosphate from solution onto porous chitin scaffolds.[89] Up to 55% by mass of calcium was deposited onto the chitin scaffold and this approach could be a useful method for the preparation of materials containing chitin and calcium for tissue engineering. Most recently, chitosan-hydroxyapatite nanocomposites have been prepared from a chitosan solution combined with a phosphoric acid and added dropwise to a $Ca(OH)_2$ suspension.[90] The resultant slurry was post treated to give the final composite. The composite was mechanically flexible and promoted bone formation.

2.4.2 Chitin-Based Bone Substitute Patents

Chitin patents belonging to this category usually find chitin applied as a carrier component in combination with calcium containing materials. A setting material comprising calcium phosphate powder in combination with an acidic polysaccharide solution, that may include chitosan and/or carboxymethyl-chitin, to give a gum-like setting material has been described.[91] The likely purpose of the chitosan was as a binding agent to the inorganic calcium, providing a mechanism whereby after setting, the resulting material had high strength. This hardened material was asserted to be useful as dental cement or bone prosthetic materials for treating bone defects and when made into block forms, as artificial bone or dental root. In another invention, osteoinducting substances derived from animal bone obtained by an elaborate extraction process have also been combined with acidic solutions of chitosan with the inclusion of calcium hydroxyapatite to give bone-filling material sheets.[92] The advantage of this invention was the non-heat or low temperature process in generating the bone-filling material. The resultant wet sheet was elastic and rubbery as prepared but is dehydrated for sterilization and regenerated to regain its elastic character by soaking in physiological saline. This osteoconducting sheet promoted bone growth in the cranium of rat animal models.

Khor et al have also patented an invention that utilizes chitin as a binding agent.[93] Finely powdered calcium hydroxyapatite was first suspended and dispersed in a 5 % lithium chloride-dimethylacetamide solvent to which chitin flakes are added. As the chitin dissolves, a viscous solution develops in which the hydroxyapatite is well dispersed. Subsequent processing maintains the well-dispersed hydroxyapatite in the solid state as films or porous sponges that are useful for bone filling as well as tissue-engineering scaffolds. Preliminary animal studies show non-toxicity of the solid materials. A bone-forming graft that contains chitin was the subject of another invention.[94] Several known biodegradable materials can be combined in solution and after cooling poured onto a gelatin sponge of defined shape that has a continuous porous framework (from the inside to the outside). After freeze-drying to form a composite-like entity, bone morphogenetic protein (BMP) is added to give the bone-forming graft. Rat animal model studies showed good bone forming with attendant resorption of the graft.

Chitosan has also been used as a binding agent for hard tissue stimulating agents primarily glycosaminoglycans, utilizing chitosan's polycationic property to combine with anionic glycosaminoglycans, for example heparin, where interaction can be ionic or covalent through a chemical reaction.[95] In a typical example, chitosan is coated onto titanium screws used in restorative dentistry and combined with heparin, washed and sterilized. In animal studies using adult rabbits, prominent bone formation was observed.

The preceding examples lead to the conclusion that chitin and chitosan are used normally as "binding" agents in orthopedic/dental applications. As has been shown, chitin, chitosan and their derivatives promote bone regeneration and future refinement could give rise to useful bone substitute materials.

2.5 TISSUE ENGINEERING

In the late 1980's, a new trend to repair or replace damaged tissue and organs began to emerge. This approach encompassed by the term Tissue Engineering (TE) is defined as "an interdisciplinary field that applies the principles of engineering and the life sciences toward the development of biological substitutes that restore, maintain and improve the function of damaged tissues and organs".[96] One of the strategies in Tissue Engineering is the use of biodegradable polymers to form a porous matrix or scaffold onto which cells are seeded. In time, the cells proliferate the scaffold to form a "tissue system". In the ideal situation, the "tissue system" after transplantation into the body becomes integrated with the host tissue as the scaffold gradually biodegrades. The encapsulation of bioactive substances such as pancreatic β cells that can secrete insulin at a controlled rate is another major component in Tissue Engineering that uses biodegradable polymers.

Chitosan is among one of the many candidates suitable as a biodegradable polymer to form scaffolds in Tissue Engineering.[97] It is readily fabricated into various shapes and sizes, processed into fiber, knitted and weaved. This provides the capability of pre-fabricating the scaffold in the shape of desired tissues or organs that can include 3-D scaffold structures. Chitosan is also insoluble at the physiological pH of 7 and therefore maintains its structure once formed. Furthermore, it's monomeric constituent is similar to the extracellular matrix environment of humans, and when biodegraded, should generate non-toxic, non-harmful residues. Finally, the prospect to chemically modify chitosan at its C-6 and N-2 positions to impart desired features offers great flexibility to this biopolymer.

Controlled freezing followed by lyophilization of chitosan solutions and gels is the general method to fabricate porous chitosan scaffolds. The benefit of this process is scaffolds having regulated micro dimensions in various shapes, namely, bulk scaffolds, porous microcarriers or beads and tubular chitosan scaffolds. Freezing establishes the nucleation of ice crystals that grow along the lines of thermal gradients. By varying the freezing conditions, the pore size is modulated from 1-250μm, dimensions relevant for cell attachment and proliferation as cell types dictate different pore sizes. Subsequent lyophilization removes the ice crystals to generate the chitosan scaffolds.

Figure 2.4 shows the SEM (scanning electron microscope) photomicrograph of the porous nature of the tubular scaffold. Porous scaffolds display mechanical behavior usually associated with composite materials where a low-modulus region at low strains and a high-modulus region at high strains, dependent on the pore size and orientation.

As an alternative to chitosan, the preparation of chitin scaffolds has also been achieved using similar strategies of freezing and lyophilization.[98] Matrices with pore sizes ranging from <10μm to 500μm were fabricated, again a function of freezing temperature and chitin gel density. In an effort to break the 500μm limit as well as to create an open-pore architecture, a novel chemical method obviating lyophilization has been developed.[99] In this process, calcium carbonate was included in the precursor chitin solution that upon gelling was reacted with dilute HCl to give defined homogeneous open-pore systems of 100-500μm and 500-1000μm pore size with porosities of ~76% and 81% respectively.

The utility of chitosan matrices as TE scaffolds has been studied with a number of cell types. Cartilage regeneration has been one key focus as it is a difficult tissue to replace.[100] Particularly promising has been the use of the cationic property of chitosan as the basis to forming insoluble complexes with chondroitin-4-sulfate-A (CSA) to fabricate membranes for the growth of bovine articular chondrocytes.[101] The chitosan-CSA hydrogel supported the maintenance of the articular chondrocyte phenotype expression in morphology and mitosis.

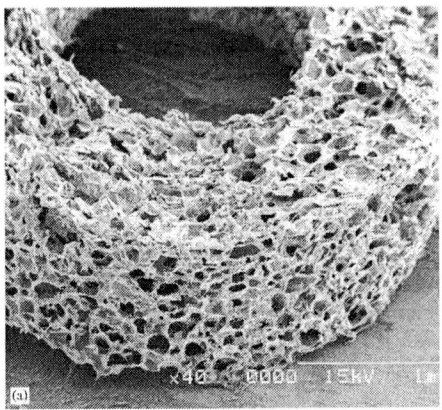

Figure 2.4   The porous character of tubular chitosan scaffold
"Reprinted from Biomaterials, Volume 20, Sundararajan V. Madihally,
Howard W.T. Matthew, Porous Chitin Scaffolds for Tissue Engineering,
pp1133-1142, 1999, with permission from Elsevier Science".

The authors suggest that the membrane could be useful as a carrier for autologous chondroctyes or as scaffolds for the generation of cartilage-like "tissue systems". In a separate study, Frondoza et al using only chitosan, obtained similar results with osteoblasts and chondrocytes.[102] When chitosan was placed in contact with human osteoblasts and chondrocytes, the osteoblasts continued to express type I collagen and chondrocytes expressed type II collagen, indicating that chitosan has elicitor-like properties for the expression of extracellular matrix (ECM) proteins in human cells, and potential for the bio-engineered repair of cartilage and bone defects.

Finally, a bi-layer chitosan film-sponge has been produced that supports the growth and proliferation of human neofetal dermal fibroblast.[103] A chitosan film is first obtained by solution casting followed by drying onto which a second solution of chitosan, containing a porogen such as NaCl or sucrose, is poured on top and subsequently freeze-dried. Upon soaking the combined freeze-dried chitosan layers, the porogen dissolves to give a porous chitosan sponge attached to the chitosan film. This chitosan film-sponge was used as a substrate onto which human neofetal dermal fibroblast cells were seeded. The cells bounded tightly to the chitosan sponge and the bilayered chitosan scaffold was proposed as a potential TE skin substitute.

Perhaps one of the first patents that can be claimed to be a forerunner of the use of chitin to the field of Tissue Engineering was the description of a method to grow cells in a 3-D pattern using a non-protein matrix.[104] Employing chitosan as a component of the liquid media or as a solid substrate, the growth of myocytes was achieved with the exclusion of unwanted fibroblasts, tumor cells and mycoplasma.

Chitosan has the ability to form porous gel matrices that can immobilize and distribute cells. This was the basis for the invention that described the encapsulation of PC12 cells used for treating Parkinson's disease. Chitosan participates as a gel matrix for the PC12 cells that are placed in an immunoisolator system that protects the PC12 cells from the body's immune system while permitting the in-flow of nutrients and outflow of clinically relevant chemicals.[105] The cells release several factors besides dopamine that provide an overall quality of treatment of the disease. In an extension to this invention, a particulate non-crosslinked preparation to entrap living cells in a core-matrix surrounded by a semipermeable membrane using chitosan has also been describe.[106] Again the membrane immunologically isolates the cells while chitosan in this instance holds and disperse the cells evenly in the inner core. Finally, a patent on the maintenance of cells in an artificial reactor completes the examples on the emerging importance of Tissue Engineering. In this patent, hepatic cells were retained in a collagen-chitosan matrix.[107]

The utilization of chitosan in Tissue Engineering has just begun. Those that transform chitin into various forms of scaffolds will also be forthcoming and the future may include oligomeric and derivatized forms of chitin as vehicles to organize cells or even tissue in the Tissue Engineering endeavor.

2.6 DRUG DELIVERY

Drug delivery is concerned with the combining of drugs with other constituents to provide dosage forms suitable for administration to the patient. The non-drug constituents serve roles such as bioprotection of the drug or the body from the drug and absorption enhancement of the drug. From a more recent perspective, emphasis has been placed on controlled release, sustained release and site-specific delivery, the delivering of a regular drug dosage for an extended period of time at the target site to attain a beneficial therapeutic effect. In addition, the lines have become grayer as the term has been expanded to include biologics in which cells are encapsulated for delivering physiologically active substances, bordering on Tissue Engineering, as discussed in the previous section. Regardless, the primary components are an active agent i.e. drug or biologic combined with a polymeric material. The polymer normally deteriorates *in vivo*, preferably at a constant rate, releasing the drug or functions as a semi-permeable membrane in the release of the active agent. Common requirements for the polymeric material are compatibility with the active agent, non-toxicity, stability, sterilizability and biodegradability. An assessment of these factors identifies chitin as a candidate that fulfills the basic requirements.

2.6.1 Chitin-Based Drug Delivery Systems

Chitin and its derivatives, predominantly chitosan, have a well-established presence in the drug delivery scene. The ready solubility of chitosan in dilute aqueous acids eases its utilization and is the rationale for its popularity over chitin. The numerous routes of entry and forms that have been conceived demonstrate the utility of chitosan as a drug delivery vehicle where importance has been placed on chitosan enhancing drug absorption, controlled release and bioadhesive properties.[108] The most popular method of administration by far is oral where microparticulate, liposomal, buccal disk, solution, vesicle, film coated, tablet and capsule forms are known. In parenteral delivery, microspheres and solutions are the most common while solutions have also been used for nasal delivery. Suspensions in ocular, encapsulation in gene therapy and gel systems completes the list of methods that have been described. Outlined below are several illustrative methods employing the biopolymer as a component in the preparation of drug delivery systems.

2.6.1.1 General Forms

Some early investigations of chitin as possible drug delivery vehicles involve the preparation of chitin and chitosan gels. In a typical preparation, chitin was first dissolved in hexafluro-2-propanol and a solubilized drug was added. Subsequent evaporation of solvent and drying gave the drug containing chitin gel. Dried chitosan gels were similarly obtained except the solvent used was aqueous acetic acid.[109] Sustained release of indomethacin and papaverine hydrochloride was studied, with the authors concluding a possible role for chitin and chitosan as sustained release agents. Similar preparation methods were also used to incorporate steroids, specifically β-oestradiol, progesterone and testosterone, into chitosan films and beads.[110] Release studies indicated gradual diffusion of steroids from the beads showing near zero-order release. Dissolution studies of the chitosan films and beads indicate no degradation over 30 days, suggesting their usefulness as controlled release systems limited perhaps only in the instance when the drug is of a macromolecular nature, where degradation of the carrier is necessary to effect drug release.

The influence of film type and thickness on drug release was studied by Kanke et al involving a drug loaded chitosan monolayer (ML) film, a double layered (DL) film comprising of an ML film plus a drug free ML film and a double layered DL-N film where the chitosan in the drug containing ML film was N-acetylated.[111] *In vitro* release studies showed that the best sustained-release profile was obtained with the DL-N film.

Chitin and chitosan have also been used as tableting agents to enhance the dissolution properties of poorly soluble drugs.[112] Chitosan and hydroxypropyl-chitosan formulations were also developed for implantable sustained anticancer drug delivery systems suitable for zero-order drug release.[113] In one preparation, uracil with chitosan and hydroxypropyl-chitosan powders were mixed; dissolution ensued with the addition of water followed by acetic acid. The resulting solution was cast and subsequent drying gave a membrane. Stick forms were obtained by extruding uracil containing chitosan solution through a nozzle into dry air. Acid neutralization was achieved by exposure to ammonia gas. *In vitro* and *in vivo* studies indicated sustained release of uracil suggesting the suitability of these formulations.

Another approach to the use of hydroxypropyl-chitosan was as a pro-drug, the chemical attachment of a drug to the biopolymer with subsequent dissociation of drug *in vivo*. The reaction of the amino functionality of chitosan with the antitumor agent cis-diamino-dichloroplatinum (CDDP) was one example.[114] Fiber forms prepared as a cotton mesh containing CDDP were implanted into mice on the tumor surface. Results indicated this method of drug delivery provided antitumor efficacy with no nephrotoxic side effects normally encountered with systemic delivery. The pro-drug method was also adapted for the carminomycin-chitosan system using a dialdehyde conjugation process.[115] Chitosan has also been used to control the release of basic fibroblast growth factors (bFGF).[116] An aqueous solution of hydroxypropylchitosan containing the bFGF was impregnated into Gore Tex® vascular grafts essentially entrapping the growth factors as the biopolymer deposits in the pores of the graft. Sustained release of bFGF by the slow dissolution of hydroxypropylchitosan was attained, suggesting another possible use of chitosan.

Tokura et al have summarized the possibility of carboxymethyl-chitin (CM-chitin) as a candidate for drug delivery in pro-drug platforms, adsorption and entrapment material or a combination of adsorption and entrapment[117] For example Watanabe et al incorporated 30% doxorubicin in a CM-chitin gel and demonstrated the time-dependent release of the drug from the gel using a lysozyme degradation system.[118] Chitosan films containing diazepam have also been prepared for oral administration.[119] Results obtained from rabbit model studies indicated that film formulations were a suitable alternative to the tablet form.

The interaction with other polymeric systems is another avenue to illustrate the versatility of chitin. One example was the simple mixing of chitosan, sodium alginate and drug powders followed by compaction into a tablet form.[120] The formulation was found to have good bioadhesion to the buccal and sublingual mucosa and suitable for intraoral drug delivery. Semi-interpenetrating polymer network (semi-IPN) membranes have been developed for possible pH sensitive drug delivery purposes.[121] *In vitro* results indicated the gels to be sensitive to simulated gastric fluid of low pH, swelling significantly, whereas in simulated intestinal fluid at neutral pH, the swelling was nominal. Chandy et al have also demonstrated release of heparin and aspirin from a poly(ethylene-vinylacetate) (PE-VAc) copolymer-chitosan co-matrix.[122] Aspirin was loaded onto chitosan beads that was in turn placed in a PE-VAc solution and the solvent allowed to evaporate to generate the co-matrix. Release studies indicated a burst effect at the start that could be attenuated by an additional polystyrene butadiene barrier membrane. The authors suggested a possible use of this system for drug delivery but recommended further confirmation studies. Chitosan-gelatin sponges are another system that has been shown to give controlled release prednisolone.[123] Finally, a new pH-sensitive system utilizing an inorganic material tetra-ethyl-orthosilicate (TEOS) with chitosan has been reported.[124] In this preparation, TEOS formed the network structure into which chitosan is impregnated to form the transparent IPN membrane (Figure 2.5). The membrane swells at low pH and shrinks at physiological pH and is useful for bioseparation. When loaded with a drug, it can be used as a delivery system.

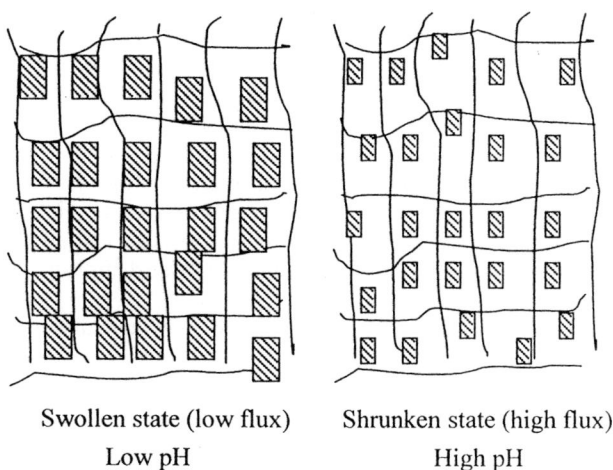

Swollen state (low flux)   Shrunken state (high flux)

Low pH   High pH

Figure 2.5:   TEOS-chitosan interpenetrating network
(line represents TEOS while box represents chitosan)

"Reprinted from Biomaterials, Volume 22, S-B. Park, J-O. You, H-Y. Park, S.J. Haam, W-S. Kim, A Novel pH-Sensitive Membrane from Chitosan-TEOS IPN: Preparation and its Drug Permeation Characteristics, pp323-330, 2001, with permission from Elsevier Science".

2.6.1.2 Microspheres

Microspheres and their more recent successor nanospheres, are a popular method of effecting drug delivery systems useful in parenteral applications. The primary effort has been directed at better drug release profiles or the protection of body tissue from harmful side effects of the drug. Chitosan microspheres incorporating cis-diamino-dichloroplatinum (CDDP) was an early example.[125] In a series of papers, Nishioka et al described the simple preparation by dispersing the drug in chitosan solution followed by vortexing until the desired sized microspheres were attained. Subsequent crosslinking was achieved with glutaraldehyde to consolidate the microspheres followed by lyophilization. In their studies, the authors found that by increasing the amount of chitosan or adding chitin to the mixture prior to forming the microspheres, the quantity of CDDP incorporated also increased while the release rate of CDDP became more regulated eliminating the initial burst effect. The improvement in performance was attributed to the enhanced solidification arising from the higher amount of biopolymers, giving microspheres with better integrity.

Reduction of the harmful side effects on the gastric mucosa has been the purpose for the encapsulation of the drug diclofenac sodium (DS) as chitosan microspheres.[126] Prepared by dispersing the drug in a chitosan solution followed by dropping via a syringe/needle assembly into a non-solvent, tripolyphosphate, the resulting microspheres were roughly 750μm in diameter with a very narrow size distribution. A slow release of the drug was obtained over 6 hours *in vitro. In vivo* studies using a rabbit model indicated effective protection of the gastric mucosa. An updated version of this work used a NaOH-methanol mixture as the non-solvent.[127] Other drugs that have been incorporated using similar techniques include 5-fluorouracil, where the drug release characteristics were again modified with other substances such as alginic acid, chitin and agar; bisphosphonates for treating pathological bone conditions and gadolinium-DTPA for neutron capture cancer therapy.[128] Microparticles of chitosan have also been investigated for use in oral vaccination.[129] The microparticles were first prepared from a chitosan solution with a surfactant. After stirring and sonication, the chitosan solution was subsequently loaded with a model antigen, ovalbumin. The experimental results indicated good uptake of ovalbumin and release studies using Balb/c mice showed good uptake by Peyer's patches suggesting potential as a vaccine delivery system. The versatility in application of chitosan microspheres has also been extended as nasal administration systems.[130] In this alternative delivery route, chitosan, containing the luteinizing releasing hormone (LHRH) agonist used in prostate and advanced breast cancer treatments, have been prepared and LHRH bioavailability was demonstrated.

The delivery of genetic materials for the treatment of hereditary diseases is an application where the use of polycationic systems bound to DNA via ionic interactions have been proposed. Chitosan was one of the candidate materials. In their study on DNA-polycation nanospheres, Leong et al reported the possible benefit of using chitosan as a delivery system.[131] Among the findings with chitosan was the ready formation of nanospheres, the possibility of conjugating ligands to the nanospheres for targeting, inclusion of other bioactive agents, protecting the DNA during transit and lyophilization of the nanospheres for storage. In several known reports, modified chitosan was typically utilized a to enhance recognition of the cellular target. Murata et al have synthesized chitosan containing pendant galactose, suited for this purpose.[132] Starting with N-trimethyl-chitosan (TM-chitosan), galactose units were attached to the C-6 position of the sugar-monomer unit. The TM-chitosan was demonstrated to have low cytotoxicity; the TM-chitosan-galactose unit conjugated with DNA demonstrated receptor recognition. Subsequently a tetragalactose group was attached that was shown to improve the target recognition ability.[133] Another substituent that has been synthesized is deoxycholic acid.[134] The attachment was achieved by activating carboxylic functionality of deoxycholic acid with a carbodiimide reagent followed by attachment to the amino group of chitosan. Approximately 5 attachments per 100 glucose units was attained that was sufficient to form self-aggregates in phosphate buffered saline and demonstrated effective transfection.

As the increase in awareness of chitin grew, the usage of chemical derivatives was inevitable. An example is the complex coacervation of carboxymethyl-chitin aqueous solution, an anionic derivative, containing 6-mercaptopurine, an anticancer agent, with iron(III) chloride solution.[135] The iron(III) ions interact with the anionic carboxymethyl group in an ionic crosslinking mechanism consolidating into microspheres upon dehydration. Genta et al utilizing a spray-drying technique prepared ampicillin containing methylpyrrolidinone-chitosan microspheres.[136] The drug release in this instance was very quick and the system was not recommended as useful for the purpose of extended term drug delivery. However, the utility of the spray-drying method was satisfactory and follow up studies using chitosan microspheres found a possible application in wound treatment.[137]

Microspheres of poly(ethylene oxide)-modified chitosan have also been developed for blood contact applications.[138] The role of the poly(ethylene oxide), derived from a commercial triblock copolymer, was to provide the blood compatibility component while chitosan functions as a cationic adsorbent for the removal of toxic compounds such as bilirubin in blood, by an extracorporeal circuit. Muzzarelli et al have also introduced oxychitin-chitosan microcapsules as a possible alternative to alginate anionic systems.[139] Finally, the possibility of using polyethylene glycol (PEG) grafted onto chitosan nanoparticles as peptide drug carriers are being explored.[140] The formation of nanoparticles was dependent on the PEG content in the range of 32 to 53%, with the 53% PEG-g-chitosan taking up the most insulin while releasing the peptide readily.

2.6.2 Chitin-Based Drug Delivery Patents

Chitin patents pertaining to drug delivery again follow roughly the pattern of scientific development, the primary focus being controlled release, enhancing bioabsorption and imparting bioprotection.

Naturally, the early work focused on using chitin as a tableting component, for example the use of chitosan as a durable and competent tablet disintegrating aid.[141] In a formulation containing 2 to 20% chitin/chitosan with water-soluble or slightly water-soluble drugs, tablets produced by compression or wet granulation disintegrated readily. Chitosan combined with angiotensin-converting enzyme(ACE) inhibitors for the treatment of hypertension was another tablet formulation.[142] In this invention, chitosan and ACE powders are finely mixed and compressed to give the tablet that converts to a gel when in contact with body fluids releasing ACE uniformly over an 8-hour period. Another example of a tablet form was the tri-coat formulation designed for drug survival until the large intestines are reached, upon which the drug is released.[143] A drug core was coated with an intermediate layer of chitosan and an outermost layer of an enteric-soluble polymer. Other inventions intended for such infragastrointestinal delivery have also been described.[144]

There are many general methods for making capsules, for encapsulating biological materials or for drug delivery purposes. Daly et al describe an example where chitosan-alginate capsules of approximately 6mm diameter were prepared.[145] Chitosan solution was dropped into alginate solution and the capsule is formed due to the complex coacervation that occurs between two oppositely charged polymers. It is conceivable that biological material or drug introduced into the chitosan solution would be encapsulated in the process. Another microcapsule preparation starts with an oil-in-water emulsion containing the active agent. After homogenizing, chitosan solution is dropped into the emulsion forming emulsion droplets. Desolubilizing chitosan around the emulsion droplets followed with lyophilization gives the microcapsule.[146] An interesting variant is the positively charged chitosan microspheres or powders. The zeta potential is kept between +0.5 to +50 mV providing prolonged absorption of the included active agents across the mucosal tissue.[147]

Low molecular weight chitosan has also been exploited to augment the availability of sparingly soluble drugs.[148] By mixing powders of the drug low molecular weight chitosan with a little water to permit kneading, a solid dispersion is obtained upon drying. *In vitro* results indicate an enhanced drug presence in solution from theses dispersions.

The use of chitosan's amino group as a chelate has also been exploited.[149] Sulfated chitosan has been combined with iron(III) for release in the gastrointestinal tract. In patient trials, the amount of iron absorbed by the human body was found to be much higher in all subjects. Ionic complexation is another invention that utilizes chitosan's chemical functionality, in this instance its polycationic nature.[150] Low molecular weight or depolymerized chitosan is first prepared and is subsequently complexed to anionic pharmaceutically active agents. This preparation can be used in tablet form but preferably as a film coating of an implant device. When in contact with body fluids, the drug slowly dissolves achieving its controlled release design.

Porous forms of chitosan have also been exploited. In one invention, chitosan solution was dropped into a coagulation bath to generate granules.[151] The gel-like granules are next introduced into a soaking bath containing a drug solution, for example an anticancer drug that is imbibed by the gel-granules. After drying, the drug-loaded granules release the anticancer agent upon the biodegradation of chitosan *in vivo*. A similar approach is that described by Kifune where a chitin-dope that was coagulated to give a swelled membrane, imbibed the drug from solution, producing a drug donor on drying. As the chitin degraded, the drug is released.[152] Kifune also invented another biodegradation system for the anticancer drug cisplatin.[153] Chitin solution was first made into fibers followed by partial deacetylation to around 40%. The fiber was next soaked in a saline solution containing cisplatin where chemical action with the free amino groups chemically bonds the drug. *In vivo*, lysozyme degrades chitin, releasing cisplatin over a 4-month period. Biodegradability was also exploited in a preparation intended for periodontal use in which the drug is encapsulated in an alginate-chitosan microspheres.[154]

Another method to prepare porous chitosan matrix was by casting a drug-loaded chitosan solution into a film.[155] The dried film is milled to give small particles releasing the drug over an extended period. A further variation involved fibers made from drug-loaded solutions that accomplished a slow release of the active agent.[156]

An interesting microsphere invention was the incorporation of magnetic particles that could be used to localize the microspheres at specific sites, for example the blood-brain region, using an external magnetic field.[157] The microspheres were made from a chitosan solution containing a ferrofluid suspension. Drug was either incorporated during the microsphere preparation or attached externally. A cell-receptor glycosaminoglycan was attached to the microsphere for binding to endothelial cells.

Solutions have also been used as delivery forms. For example insulin was combined with chitosan for enhanced absorption either as a solution or microspheres. A solution cocktail of morphine-6-glucuronide with chitosan used as spray mist in nasal administration has also been patented.[158] Another patent combined chitosan with influenza vaccine in a 1:1 solution for intranasal administration. Chitosan was indicated as a useful adjuvant to stimulate the mucosal immune system thereby increasing the effectiveness of the vaccine.[159] Last, the interpenetrating network by Amiji with chitosan and polyethylene-oxide described in the scientific section was also patented.[160]

In summary, many ways of using chitin, chitosan and their derivatives have been described in the literature for drug delivery. While some forms such as tablets, films and gels do not require much sophistication in formulation, others such as the microspheres intended for internal use would require proper fabrication that takes biodegradability into consideration. The present emphasis is on efforts to obtain a better understanding of the modulating effects of chitosan on drug transports, elucidating mechanisms such as mucoadhesion and action at the tight epithelial cells junction. As these issues become better resolved the vast potential for chitin in the drug delivery field will increase.

2.7 RELEVANCE OF CHITIN BIOMEDICAL APPLICATIONS

It can be surmised from the overview survey that at least from a scientific perspective, opportunities for chitin in biomedical applications exists. How should this standpoint be transformed into action that can bring about the fruition of chitin based biomedical products? Is it realistic to attempt? Two questions have to be answered, the market relevance and chitin's relevance in this market. The five focus areas in the preceding discussion were wound dressings; blood interactions; orthopedic implants; tissue engineering and drug delivery. To assess the market relevance, we take a look at the global market in each area. The possibility for chitin to penetrate the respective biomedical markets has to be considered from the segment's materials requirements and the potential match with chitin's properties.

The Global market for wound management products has been estimated at US\$10 billion with an expected annual growth rate of 12% to reach US\$15 billion in the first decade of the 21st Century.[161] The market covers primarily wound cleansing and debridement products, growth factors and skin replacements. The US market alone commands an estimated US\$5 to US\$7 billion for the treatment of chronic wounds that include venous stasis ulcers and diabetic foot ulcers and is set to experience tremendous growth close to 10% over the next few years. Growth factors and skin replacements are the highest-value end products.

The ideal wound dressing (whether necessary or desired) that can address all stages in the wound healing process, incorporating features such as infection control, exudates absorption and skin replacement, is still unmet. At one end of the spectrum, cellulose-based gauze predominates as a cost effective treatment. However, the adhesion to wounds and exposure to infection and moisture loss subjects this method to limitations that at best are tolerated. At the other end of the spectrum, effective skin replacements include polyurethane films, animal based tissue substitutes and Tissue Engineered artificial skins have been demonstrated to produce good wound healing and recovery. Again, each has their limitation and with the increasing concern regarding the transmission of animal-borne diseases, the use of animal derived biomaterials may be further limited in the future, and in the case of tissue engineered products, treatment is obtained at a high cost, in some instances upwards of US$2000 per square foot. This is an area that has vast potential and where chitin can play an important role.

The primary requirements for a wound dressing are in providing a barrier to prevent rapid moisture loss as well as infection from microorganisms. Outlined in section 2.2 were many methods that have been devised to use chitin and chitosan in film and membrane forms that can qualify the barrier condition. More importantly, chitin and chitosan have been shown to promote wound healing. Several studies indicate that chitin affects the wound bed by activating factors such as the PMNs and MMPs that accelerate the recovery of the wound. Further studies are required on the role N-acetyl-glucosamine plays and whether degradation by the enzyme lysozyme present in wounds plays a role in activating wound healing. Regardless, it must be noted that in wound dressing applications, chitin will probably participate as a component of an external medical device. If utilized solely in a barrier role, it is unlikely to make an impact in the wound market where established products abound. Inclusion of growth factors and proteolytic enzymes in chitin-based wound dressings would enhance the attractiveness of using chitin. Finally, as the author had suggested earlier, based on the present understanding in the science of wound healing and management, the preferred chitin embodiment in the wound dressing market is as an artificial skin with the exudate absorbing and cell seeding capability present. On the scientific evidence, chitin can be advanced as a player in this arena, both as a dressing in its own right as well as a scaffolding material in place of presently used materials such as polylactic acid and nylon.

In blood interaction applications, the hemostasis market was US$250 million in 1999, and expected to grow at a modest annual rate of 1.6% to reach US$270 million in 2005.[162] Chitosan is a potential player in this market. Recently, a poly-N-acetyl-glucosamine based product has been reported to been given US-FDA approval for sale.[163]

However, the more lucrative market appears to be the application of the anticoagulant property of chitin materials. There are many clinical procedures that require blood to flow through polymer tubing such as in kidney dialysis and open-heart surgery with a combined market value of more than US$20 billion annually. Other blood interaction situations include peritoneal dialysis, blood bags and catheters. While new technologies such as the beating-heart procedure can reduce the number of open-heart surgeries, complete replacement is not imminent yet. Presently, the prevention of blood coagulation and related complications such as thrombosis and hemolysis to the patient are arrested by the systemic administration of the anticoagulant, heparin. While heparin is effective, side effects to the patient include the risk of slower recovery post-operatively and uncontrolled bleeding in the event of a procedural accident. Therefore, a reduction or elimination in the heparin dosage administered is favored. Towards this goal, many alternatives have been investigated.

For example hemocompatible materials such as polyurethane have been used as the outer coat of catheters to prevent blood clots during angioplasty procedures. However, the extension of polyurethane to blood contact tubing has not occurred possibly because of the cost involved. Therefore, the established approach in attempting to make blood contact tubing hemocompatible is to coat the inner lumen with anticoagulants especially the immobilization, bonding and deposition of heparin that has met with varying successes. Developing an effective hemocompatible coating for the inner lumen of blood contact tubing will be very profitable and definitely within the capability of chitin. Finally, the concept of a chitin derivative as a viable heparin analog should also be considered seriously.

The orthopedic sector is primarily a US$5 billion implant market dominated by hip and spinal implants and knee-joint replacements, where metals will continue to dominate for the foreseeable future, particularly stainless steel and titanium which are the predominant materials. The likely application of any chitin-based orthopedic material is in the sector of biodegradable screws, pins and plates, a market presently dominated by polylactic acid (PLA) and polyglycolic acid (PGA) and their copolymers that are synthetic analogs of molecules present in the body. While this market estimated to be around US$500 million might be comparatively modest, chitin has been demonstrated to promote bone regeneration. Its combination with calcium compounds and ready biodegradability warrants serious consideration of chitin in this field.

For tissue engineering, the market is just beginning. Apart from skin tissue engineering products already introduced as noted above, orthopedic products appear to be the most promising in that bone grafts and autologous chondrocyte implants are expected to compete as the main revenue generator. The market is expected to take off after 2005. The bone regeneration and biodegradability properties of chitin and its ability to be shaped into various tissue-relevant configurations are features that make chitin a contender in this highly sophisticated field of medical devices.

The drug delivery scene is the most lucrative judging by the projected billions in revenue listed in Table 2.2. For chitin-based systems, chitosan predominates this area of influence with some chitosan containing delivery systems in advanced clinical trials. However, it should be noted that the amount of material required in this application is the least among the five areas surveyed with an attendant higher price tag to the source material. On a brighter note, the capabilities of chemical derivatives synthesized in the last ten years make it possible to extend the application of chitin-based materials in drug delivery yet to be realized.

Form of delivery	2000	2005 (projection)
Controlled release	$14 billion	$26.3 billion
Injectable/implantable polymer systems	$3.8 billion	$7.2 billion
Transdermal delivery	$6.7 billion	$12.7 billion
Transnasal delivery	$8.2 billion	$16 billion
Pulmonary	$11.7 billion	$22.6 billion
Transmucosal	$2.4 billion	$6.5 billion
Rectal	$500 million	$1.2 billion
Cell/gene therapy	Nil	$5 billion
Miscellaneous	$1.5 billion	$2.5 billion
Total	$50.4 billion	$104.3 billion

Table 2.2:   Global drug delivery market in 2000 and 2005

In summary, the payback in revenue terms for biomedical applications is high. What is at stake is a high value-add market where a waste material can be transformed into products that has the potential not only to meet the mission requirements of the various biomedical applications but also offers a financial justification as well. Even as the excitement for chitin as a biomedical material has been generated, it's popularity in clinical applications has so far been in restricted markets. In more regulated markets, orally administered forms appear to have been the only penetration so far. How can the chitin potential in biomedical applications be sewn up to be readily acceptable?

It is interesting to note that at a round table discussion at the 5th International Chitin Chitosan Conference in 1992, several questions were posed to attendees.[164] The summary that was reported identified the following:

1. "A requirement for set standards to identify and specify all chito-materials.

2. A need for concise yet systematic approval process for chito-materials.

3. A need for standard analytical methods and procedures accepted by all.

4. A need for a comprehensive database for all to access."

In other words the questions have been in the minds of the chitin scientific community for some time. Certainly the science of chitin to date can already address the questions generated in 1992. For chitin to bridge the final gap necessitates several bold initiatives, some of which, to the knowledge of the author, are in progress. Bold because a belief that chitin will make it as a biomedical material comes at high capital cost risks! How can this risk be justified? What are the obstacles?

For one, a major stumbling block to the widespread biomedical use of chitin is regulatory approval. This can only occur if the production of raw chitin material, including its derivatives and the manufacturing of chitin containing medical products comply with standards. These standards can only be established if a consensus is reached on the analytical methods and procedures. Therein lies the opportunity in establishing chitin as a biomaterial.

2.8 REFERENCES

1 S.C. Gad, Safety Evaluation of Medical Devices, Marcel Dekker, Inc., New York, N.Y. 1997. 2

2 F.H. Silver, Biomaterials, Medical Devices and Tissue Engineering, Chapman & Hall, London, England. 1994. Chapter 2

3 J.M. Howell, Current and future trends in wound healing. Emergency Medicine Clinics of North America 10(4) (1992) 655-663

4 J.F. Prudden, P. Migel, P. Hanson, L. Freidrich, L. Balassa, The discovery of a potent pure chemical wound-healing accelerator. The American J. Surgery 119 (1970) 560-564

5 G.G. Allan, L.C. Altman, R.E. Bensinger, D.K. Ghosh, Y. Hirabayashi, A.N. Neogi, S. Neogi, Biomedical applications of chitin and chitosan. in Chitin, Chitosan and Related Enzymes, J.P. Zikakis, ed., Academic Press Inc., Orlando, FL, USA, 1984. 119-133

6   Y. Oshshima, K. Nishino, Y. Yonekura, S. Kishimoto, S. Wakabayashi, Clinical applications of chitin non-woven fabric as wound dressing. European J. Plastic Surgery 10 (1987) 66-69

7   C-H Su, C-S Sun, S-W Juan, C-H Hu, W-T Ke, M-T Sheu, Fungal mycelia as the source of chitin and polysaccharides and their applications as skin substitutes. Biomaterials 18 (1997) 1169-1174

8   Y-W. Cho, Y-N. Cho, S-H. Chung, G. Yoo, S-W. Ko, Water-soluble chitin as a wound healing accelerator. Biomaterials 20 (1999) 2139-2145

9   N.L.B.M. Yusof, L.Y. Lim, E. Khor, Preparation and characterization of chitin beads as a wound dressing precursor. J. Biomedical Materials Research 54 (2001) 59-68

10   F-L. Mi, S-S. Shyu, Y-B. Wu, S-T. Lee, J-Y. Shyong, R-N. Huang, Fabrication and characterization of a sponge-like asymmetric chitosan membrane as a wound dressing. Biomaterials 22 (2001) 165-173

11   G. Biagini, A. Pugnaloni, A. Damadei, A. Bertani, A. Belligolli, V. Bicchiega, R. Muzzarelli, Morphological study of the capsular organization around tissue expanders coated with N-carboxybutyl chitosan. Biomaterials 12 (1991) 287-291

12   G. Biagini, A. Bertani, R. Muzzarelli, A. Damadei, G. Dibenedetto, A. Belligolli, G. Riccotti, Wound management with N- carboxybutyl chitosan. Biomaterials 12 (1991) 281-286; G. Biagini, R.A.A. Muzzarelli, R. Giardino, C. Castaldini, Biological materials for wound healing. Advances in Chitin and Chitosan, C.J. Brine, P.A. Sandford, J.P. Zikakis, eds., Elsevier Applied Science, New York, N.Y., 1992. 16-24

13   R.A.A. Muzzarelli, E. Toschi, G. Ferioli, R. Giardino, M. Fini, M. Rocca, G. Biagini, N-Carboxylbutyl chitosan and fibrin glue in cutaneous repair processes. J. Bioactive and Compatible Polymer 5 (1990) 396-411

14   O. Damour, P.Y. Gueugniaud, M. Berthin-Maghit, P. Rousselle, F. Berthod, F. Sahuc, C. Colombel, A dermal substrate made of collagen-GAG-chitosan for deep burn coverage: First clinical uses. Clinical Materials 15 (1994) 273-276

15   G. Kratz, C. Arnander, J. Swedenborg, M. Back, C. Falk, I. Gouda, O. Larm, Heparin-chitosan complexes stimulate wound healing in human skin. Scandinavian J. Plastic Reconstructive Hand Surgery 31 (1997) 119-123

16   Y.M. Lee, S.S. Kim, M.H. Park, K.W. Song, Y.K. Sung, I.K. Kang, β-Chitin -based wound dressing containing sulfurdiazine. J. Materials. Science: Materials in Medicine 11 (2000) 817-823

17   W.K. Loke, S.K. Lau, L.Y. Lim, E. Khor, K.S. Chow, Wound dressing with sustained anti-microbial capability. J. Biomedical Materials Research 53 (2000) 8-17

18   P.C. Berscht, B. Nies, A. Liebendörfer, J. Kreuter, In vitro evaluation of biocompatibility of different wound dressing materials. J. Materials. Science: Materials in Medicine 6 (1995) 201-205

19   Y. Okamoto, K. Shibazaki, S. Minami, A. Matsuhashi, S. Tanioka, Y. Shigemasa, Evaluation of chitin and chitosan on open wound healing in dogs. J. Veterinary Medicine Science 57(5) (1995) 851-854

20   P.C. Berscht, B. Nies, A. Liebendörfer, J. Kreuter, In vitro evaluation of biocompatibility of different wound dressing materials. J. Materials. Science: Materials in Medicine 6 (1995) 201-205

21   S. Minami, Y. Okamoto, S-i. Tanioka, H. Sashiwa, H. Saimoto, A. Matsuhashi, Y. Shigemasa, Effects of chitosan on open wound healing. in Carbohydrates and Carbohydrate Polymers, Analysis, Biotechnology, Modification, Antiviral, Biomedical

and Other Applications, M. Yalpani, ed., ATL Press, Mount Prospect, IL, USA, 1993. 141-152

[22] T. Nakade, H. Yokota, H. Taniyama, Y. Hori, N. Agata, T. Ikeda, H, Furusaki, Y. Yamada, Y. Uchida, A. Yuasa, M. Yamaguchi, K. Otomo, Matrix metalloproteinase (MMP) 9 induced in skin and subcutaneous tissue by implanted chitin in rats. Carbohydrate Polymers 41 (2000) 327-329

[23] L.L. Balassa, Process for facilitating wound healing with N-acetylated partially depolymerized chitin materials, US Patent 3914413, October 21st, 1975; L.L. Balassa, Process for promoting wound healing with chitin derivatives, US Patent 3911116, October 7th, 1975; L.L. Balassa, Chitin and chitin derivatives for promoting wound healing, US Patent 3903268, September 2nd, 1975

[24] W.G. Malette, H.J. Quigley Jr., Method of achieving hemostasis, inhibiting fibroplasia and promoting tissue regeneration in a tissue wound, US Patent 4532134, July 30th, 1985

[25] J. Turková, J. Stamberg, Proteolytic, dry biopolymeric composition for treatment of wounds and method of using the same, US Patent 4613502, September 23rd, 1986

[26] Koji Kifune, Yasuhiko Yamaguchi & Hiroyuki Tanae, Wound dressing, US Patent 4651725, March 24th, 1987

[27] C. J. Albisetti, J. E. Castle, Dispersion Of Chitin And Product Therefrom", US Patent 4931551, June 5th, 1990

[28] B. Sagar, P. Hamlyn, D. Wales, Wound dressing, US Patent 4960413, October 2nd, 1990

[29] B.G. Sparkes, D.G. Murray, Chitosan based wound dressing materials, US Patent 4572906, February 25th, 1986

[30] D.S. Jackson, Chitosan-glycerol-water gel, US Patent 4659700, April 21st, 1987

[31] D. T. Mosbey, Wound filling compositions, US Patent 4956350, September 11th, 1990

[32] S.F. Nielson, D.W. Kim, Water absorbent latex polymer foams containing chitosan (chitin), US Patent 5011864, April 30th, 1991

[33] D.H. Lorenz, C.C. Lee, Gels formed by the interaction of polyvinylpyrrolidone with chitosan derivatives, US Patent 5420197, May 30th, 1995

[34] A.S. Fox, A.E. Allen, Gel forming system for use as wound dressing, US Patent 5578661, November 26th, 1996

[35] D.B. Eagles, G. Bakis, A.B. Jeffery, C. Mermingis, T.H. Hagoort, Method of producing polysaccharide foams, US Patent 5840777, November 24th, 1998

[36] A.S. Pandit, Hemostatic wound dressing, US Patent 5836970, November 17th, 1998

[37] C.J. Hardy, Swellable wound dressing materials, US Patent 6022556, February 8th, 2000

[38] K. Motosugi, K. Kifune, Y. Yamaguchi, Y. Nobe, H. Tanae, Shaped chitin body, US Patent 4699135, October 13th, 1987

[39] E. Khor, A.C. Aun Wan, G.W. Hastings, Method of preparing water swellable gel from chitin, US Patent 6025479, February 15th, 2000

[40] M. Sakai, Wound dressing, US Patent 4803078, February 7th, 1989

[41] M. Koide, J. Konishi, K. Ikegami, K. Osaki, Multilayer wound covering materials comprising a supporting layer and a moisture permeation controlling layer and method for their manufacture, US Patent 5395305, March 7th, 1995

[42] H. Ohta, K. Takaishi, N. Uetani, Wound protecting member including chitin, US Patent 5405314, April 11th, 1995

43   T. Miyata, K. Kodaira, H. Higashijima, T. Kimura,Y. Noshiki, Biomaterial comprising a composite material of a chitosan derivative and collagen derivative, US Patent 5116824, May 26th, 1992

44   R. Muzzarelli, Methyl pyrrolidinone chitosan, production process and uses thereof, US Patent 5378472, January 3rd, 199

45   M. Takai, Y. Shimizu, J. Shimizu, K. Yamazaki, Y. Kumabayashi, H. Shimizu, K. Yamada, Wound healing composition using squid chitin and fish skin collagen, US Patent 5698228. December 16th, 1997

46   C. Collombel, O. Damour, C. Gagnieu, F Poinsignon, C. Echinard, J. Marichy, Biomaterials with a base of mixtures of collagen, chitosan and glycosaminoglycans preparing them and their application in human medicine, US Patent 5166187, November 24th, 1992

47   A. De Rosa, A. Rossi, P. Affaitati, Process for the preparation of iodinated biopolymers having disinfectant and cicatrizing activity, and the iodinated biopolymers obtainable thereby, US Patent 5538955, July 23rd, 1996

48   H.J. Kaessmann, K.W. An Haak, Chitosan foil for wound sealing and process for its preparation, US Patent 5597581, January 28th, 1997

49   I. Gouda, O. Larm, Method of promoting dermal wound healing with chitosan and heparin or heparin sulfate, US Patent 5902798, May 11, 1999

50   C.M. Elson, N, O-Carboxymethylchitosan for prevention of surgical adhesions, US Patent 5679658, October 21st, 1997

51   M. Fukuyama, K. Miwa, K. Ishikawa, Material for elimination or detoxification of super antigens, US Patent 5928633, July 27th, 1999

52   R. Barbucci, A. Magnani, S. Lamponi, A. Albanese, Chemistry and biology of glycosaminoglycans in blood coagulation. Polymers for Advanced Technologies 7 (1996) 675-685

53   M.L. Wolfrom, T.M. Shen, C.G. Summers, Sulfated nitrogenous polysaccharides and their anticoagulant activity. J. American Chemical Society 75 (1953) 1519

54   M.L. Wolfrom, T.M. Shen, The sulfonation of chitosan. J. American Chemical Society 81 (1959) 1764-1766

55   D.T. Warner, L.L. Coleman, Selective sulfonation of amino groups in amino alcohols. J. Organic Chemistry 23 (1958) 1133-1135

56   R.L. Whistler, M. Kosik, Anticoagulant activity of oxidized and N- and O-sulfated chitosan. Archives of Biochemistry and Biophysics 142 (1971) 106-110

57   D. Horton, E.K. Just, Preparation from chitin of $(1{\rightarrow}4)$-2-amino-2-deoxy-β-D-glucopyranuronan and its 2-sulfoamino analog having blood anticoagulant properties. Carbohydrate Research 29 (1973) 173-179

58   R.A.A. Muzzarelli, Carboxymethylated chitins and chitosans. Carbohydrate Polymers 8 (1988) 1-21

59   W. Okiei, S. Nishimura, O. Somorin, N. Nishi, S. Tokura, Inhibitory action of sulfated chitin derivatives on the hydrolytic activity of thrombin. in Chitin in Nature and Technology, R.Muzzarelli, C. Jeuniaux, G.W. Gooday, eds., Plenum Press, New York 1986. 453-460

60   K. Hagiwara, Y. Kuribayashi, H. Iwai, I. Azuma, S. Tokura, K. Ikuta, C. Ishihara, A sulfated chitin inhibits hemagglutination by *Theileria sergenti* merozoites. Carbohydrate Polymers 39 (1999) 245-248

38

61 R.A.A. Muzzarelli, F. Tanfani, M. Emanuelli, D.P. Pace, E. Chiurazzi, M. Piani, Sulfated N-(carboxymethyl)chitosans: Novel blood anticoagulants. Carbohydrate Research 126 (1984) 225-231; R.A.A. Muzzarelli, F. Tanfani, M. Emanuelli, E. Chiurazzi, M. Piani, Sulfated N-carboxymethylchitosans as blood anticoagulants. in Chitin in Nature and Technology, R.Muzzarelli, C. Jeuniaux, G.W. Gooday, eds., Plenum Press, New York, 1986. 461-467

62 W.G. Malette, H.J. Quigley Jr., R.D. Gaines, N.D. Johnson, W.G. Rainer, Chitosan: A new hemostatic. Annals of Thoracic Surgery 36 (1983) 55-58

63 W.G. Malette, H.J. Quigley Jr., Method for the therapeutic occlusion of blood vessels, US Patent 4452785, June 5th, 1984

64 G. Fradet, S. Brister, D.S. Mulder, J. Lough, B.L. Averbach, Evaluation of chitosan as a new hemostatic agent: *In vitro* and *in vivo* experiments. in Chitin in Nature and Technology, R.Muzzarelli, C. Jeuniaux, G.W. Gooday, eds., Plenum Press, New York, 1986. 443-451

65 K. Kaifu, T. Komai, Wetting characteristics and blood clotting on surfaces of acylated chitins. J. Biomedical Materials Research 16 (1982) 757-766

66 S. Hirano, Y. Noishiki, The blood compatibility of chitosan and N-acylchitosans. J. Biomedical Materials Research 19 (1985) 413-417

67 H. Fukuda, Y. Kikuchi, *In vitro* clot formation on the polyelectrolyte complexes of sodium dextran sulfate with chitosan. J. Biomedical Materials Research 12 (1978) 531-539

68 K.Y. Lee, S.H. Wan, W.H. Park, Blood compatibility and biodegradability of partially N-acylated chitosan derivatives. Biomaterials 16(16) (1995) 1211-1216

69 J. Dutkiewicz, L. Szosland, M. Kucharska, L. Judkiewicz, R. Ciszewski, Structure-bioactivity relationship of chitin derivatives-Part 1: The effect of solid chitin derivatives on blood coagulation. J. Bioactive and compatible polymers 5 (1990) 293-304

70 J. Dutkiewicz, M. Kucharska, A. Papiewski, L. Judkiewicz, R. Ciszewski, Chitosan sealant for vascular grafts with no need of heparinization. in Advances in Chitin and Chitosan, C.J. Brine, P.A. Sanford, J.P. Zikakis, eds., Elsevier Applied Science, London and New York, N.Y., 1992. 54-60

71 A. Hoekstra, H. Struszczyk, O. Kivekäs, Percutaneous microcrystalline chitosan application for sealing arterial puncture sites. Biomaterials 19 (1998) 1467-1471

72 W.J. Hammer, Antithrombogenic articles, US Patent 4326532, April 27th, 1982

73 B. Haimovich, A. Freeman, R. Greco, Thromboresistant surface treatment for biomaterials, US Patent 5578073, November 26th, 1996

74 M.M. Amiji, Permeability and blood compatibility properties of chitosan-poly-ethyleneoxide) blend membranes for hemodialysis. Biomaterials 16 (1995) 593-599

75 M.M. Amiji, Surface modification of chitosan membranes by complexation-interpenetration of anionic polysaccharides for improved blood compatibility in hemodialysis. J. Biomater. Sci. Polym. Ed. 8(4) (1996) 281-298

76 M.M. Amiji, Biocompatible articles and method for making same, US Patent 5885609, March 23, 1999.

77 J. Vincent, Structural biomaterials, Princeton University Press, Princeton, N.J., 1990. 183

78 A.M. Rouhi, Biomaterials for women. Chem. Eng. News, 77 (1999) 24-26

79   A.C.A. Wan, E. Khor, G.W. Hastings, Hydroxyapatite modified chitin as potential hard tissue substitute material. J. Biomedical Materials Research: Applied Biomaterials 38 (1997) 235-241

80   M. Maeda, H. Iwase, K. Kifune, Characteristics of chitin for orthopedic use. in Chitin, Chitosan and Related Enzymes. J.P. Zikakis, ed., Academic Press Inc., Orlando, FL, USA, 1984. 411-415

81   G. Borah, G. Scott, K. Wortham, Bone induction by chitosan in endochondral bones of the extremities. in Advances in Chitin and Chitosan, C.J. Brine, P.A. Sanford, J.P. Zikakis, eds., Elsevier Applied Science, London and New York, 1992. 47-53

82   R.A.A. Muzzarelli, V. Bicchiega, G. Biagini, A Pugnaloni. R. Rizzoli, Role of N-carboxybutyl chitosan in the repair of the meniscus. J. Bioactive and Compatible Polymers 7 (1992) 130-148

83   R.A.A. Muzzarelli, M. Mattioli-Belmonte, C. Tietz, R. Biagini, G. Ferioli, M.A. Brunelli, M. Fini, R. Giardino, P. Ilari, G. Biagini, Stimulatory effect on bone formation exerted by a modified chitosan. Biomaterials 15(13) (1994) 1075-1081

84   M. Mattioli-Belmonte, G. Biagini, R.A.A. Muzzarelli, C. Castaldini, M.G. Gandolfi, A. Krajewski, A. Ravaglioli, M. Fini, R. Giardino, Osteoinduction in the presence of chitosan-coated porous hydroxyapatite. J. Bioactive and Compatible Polymers 10 (1995) 249-257

85   R.A.A. Muzzarelli, G. Biagini, M. Mattioli-Belmonte, O. Talassi, M.G. Gandolfi, R. Solmi, S. Carraro, R. Giardino, M. Fini, N. Nicoli-Aldini, Osteoinduction by chitosan-complexed BMP: Morpho-structural responses in an osteoporotic model. J. Bioactive and Compatible Polymers 12 (1997) 321-329

86   M. Mattioli-Belmonte, N. Nicoli-Aldini, A. De Benedittis, G. Sgarbi, S. Amati, M. Fini, G. Biagini, R.A.A. Muzzarelli, Morphological study of bone regeneration in the presence of 6-oxychitin. Carbohydrate Polymers 40 (1999) 23-27

87   T. Kawakami, M. Antoh, H. Hasegawa, T. Yamagishi, M. Ito, S. Eda, Experimental study on osteoconductive properties of a chitosan-bonded hydroxyapatite self-hardening paste. Biomaterials 13(11) (1992) 759-763

88   M. Ito, T. Yamagishi, H. Yagasaki, A.H. Kafrawy, In vitro properties of a chitosan-bonded bone-filling paste: Studies on solubility of calcium phosphate compounds. J. Biomedical Materials Research 32 (1996) 95-98

89   A.C.A. Wan, E. Khor, G.W. Hastings, Preparation of a chitin-apatite composite by *in situ* precipitation onto porous chitin scaffolds. J. Biomedical Materials Research: Applied Biomaterials 41 (1998) 541-548

90   I. Yamaguchi, K. Tokuchi, H. Fukuzaki, Y. Koyama, K. Takakuda, H. Monma, J. Tanaka, Preparation and microstructure analysis of chitosan/hydroxyapatite nanocomposites. J. Biomedical Materials Research 55 (2001) 20-27

91   M. Sumita, Composition for forming calcium phosphate type setting material and process for producing setting material, US Patent 5180426, January 19th, 1993; M. Sumita, Composition for forming calcium phosphate type setting material and process for producing setting material, US Patent 5281404, January 25th, 1994

92   M. Ito, Osteoinduction substance, method of manufacturing the same, and bone filling material including the same, US Patent 5618339, April 8th, 1997

93   E. Khor, A. C.A. Wan, G. W. Hastings, Method of preparing filler containing forms of chitin, US Patent 5821285, October 13th, 1998

94   S. Yokota, S. Shimokawa, R. Sonohara, A. Okada, K. Takahashi, Bone-forming graft, US Patent 5830493, November 3rd, 1998

95   H-A. Hansson, G.J. Ruden, O. Larm, Hard tissue stimulating agent, US Patent 5894070, April 13th, 1999

96   B.E. Chaignaud, R. Langer, J.P. Vacanti, The history of tissue engineering using synthetic biodegradable polymer scaffolds and cells. In: Synthetic Biodegradable Polymer Scaffolds, A. Atala, D. Mooney, R. Langer, J.P. Vacanti, eds., Birkhauser, Boston, USA, 1997. 1

97   S.V. Madihally, H.W.T. Matthew, Porous chitosan scaffolds for tissue engineering. Biomaterials 20 (1999) 1133-1142

98   K.S. Chow, E. Khor, Fabrication of porous chitin matrices. in Advances in Chitin Science, Volume 4, M.G. Peter, A. Domard, R.A.A. Muzzarelli, eds., Universität Potsdam, Potsdam, Germany 2000. 355-360

99   K.S. Chow, E. Khor, Novel fabrication of open-pore chitin matrixes. Biomacromolecules 1 (2000) 61-67

100   J-K. F. Suh, H.W.T. Matthew, Application of chitosan-based polysaccharide biomaterials in cartilage tissue engineering: a review. Biomaterials 21 (2000) 2589-2598

101   V.F. Sechriest, Y.J. Miao, C. Niyibizi, A. Westerhausen-Larson, H.W. Matthew, C.H. Evans, F. .H. Fu, J-K. Suh, GAG-augmented polysaccharide hydrogel: A novel biocompatible and biodegradable material to support chondrogenesis. J. Biomedical Materials Research 49 (2000) 534-541

102   A. Lahiji, A. Sohrabi, D.S. Hungerford, C.G. Frondoza, Chitosan supports the expression of extracellular matrix proteins in human osteoblasts and chondrocytes. J. Biomedical Materials Research 51 (2000) 586-595

103   J. Ma, H. Wang, B. He, J. Chen, A preliminary in vitro study on the fabrication and tissue engineering applications of a novel chitosan bilayer material as a scaffold of human neofetal dermal fibroblasts. Biomaterials 22 (2001) 331-336

104   W.G. Malette, H.J. Quigley Jr., Method of altering growth and development and suppressing contamination microorganisms in cell or tissue culture, US Patent 4605623, August 12th, 1986

105   D.F. Emerich, P. Aebischer, J.H. Kordower, Encapsulated PC12 cell transplants for treatment of Parkinson's disease, US Patent 5853385, December 29th, 1998

106   P. Aebischer, B.A. Zielinski, Encapsulated PC12 cell transplants for treatment of Parkinson's disease, US Patent 58715985, February16[th], 1999

107   W.S. Hu, F.B. Cerra, S.L. Nyberg, M.T. Scholz, R.A. Shatford, Maintaining cells for an extended time by entrapment in a contracted matrix, US Patent 5981211, November 9th, 1999

108   V. Dodane, V.D. Vilivalam, Pharmaceutical applications of chitosan. Pharmaceutical Science and Technology Today, 1 (1998) 246-253

109   S. Miyazaki, K. Ishii, T. Nadai, The use of chitin and chitosan as drug carriers. Chemical and Pharmaceutical Bulletin 31 (1983) 2507-2509

110   T. Chandy, C.P. Sharma, Biodegradable chitosan matrix for the controlled release of steroids. Biomaterial Artificial Cells and Immobilzation Biotechnology 19(4) (1991) 745-760

111   M. Kanke, H. Katayama, S. Tsuzuki, H. Kuramoto, Application of chitin and chitosan to pharmaceutical preparations. I. Film preparation and in vitro evaluation. Chemical and Pharmaceutical Bulletin 37(2) (1989) 523-525

112   T. Nagai, Y. Sawayanagi, N. Nambu, Application of chitin and chitosan to pharmaceutical preparations. in Chitin, Chitosan and Related Enzymes. J.P. Zikakis, ed., Academic Press Inc., Orlando, FL, USA, 1984. 21-39

[113] Y. Machida, T. Nagai, M. Abe, T. Sannan, Use of chitosan and hydroxypropylchitosan in drug formulations to effect sustained release. Drug Design and Delivery 1 (1986) 119-130

[114] K. Suzuki, T. Nakamura, H. Matsuura, K. Kifune, R. Tsurutani, A new drug delivery system for local cancer chemotherapy using cisplatin and chitin. Anticancer Research 15 (1995) 423-426

[115] N. Todorova, M. Krysteva, K. Maneva, D. Todorov, Carminomycin-chitosan: A conjugated antitumor antibiotic. J. Bioact. Compat. Polym. 14 (1999) 178-184

[116] K. Yamamura, T. Sakurai, K. Yano, T. Nabeshima, T. Yotsuyanagi, Sustained release of basic fibroblast growth factor from the synthetic vascular prosthesis using hydroxypropylchitosan acetate. J. Biomedical Materials Research 29 (1995) 203-206

[117] S. Tokura, Y. Miura, Y. Kaneda, Y. Uraki, Drug delivery system using biodegradable carrier. in Polymeric Delivery Systems: Properties and Applications, M.A. El-Nokaly, D.M. Platt, B.A. Charpentier, eds., ACS Symposium Series 520, 1993. 351-361

[118] K. Watanabe, I. Saiki, Y. Uraki, S. Tokura, I Azuma, 6-O-Carboxymethyl-chitin (CM-chitin) as a drug carrier. Chemical and Pharmaceutical Bulletin 38(2) (1990) 506-509

[119] S. Miyazaki, H. Yamaguchi, M. Takada, W-M Hou, Y. Takeichi, H. Yasubuchi, Pharmaceutical application of biomedical polymers XXIX. Preliminary study of film dosage form prepared from chitosan for oral drug delivery. Acta Pharmceutica Nordica 2(6) (1990) 401-406

[120] S. Miyazaki, A. Nakayama, M. Oda, M. Takada, D. Attwood, Chitosan and sodium alginate based bioadhesive tablets for intraoral drug delivery. Biological and Pharmaceutical Bulletin 17(5) (1994) 745-747

[121] V.R. Patel, M.M. Amiji, pH-sensitive swelling and drug-release properties of chitosan-poly(ethylene oxide) semi-interpenetrating polymer network. ACS Symposium Series 627 (1996) 209-220

[122] S.C. Vasudev, T. Chandy, C.P. Sharma, Development of chitosan/polyethylene vinyl acetate co-matrix: controlled release of aspirin-heparin for preventing cardiovascular thrombosis. Biomaterials 18(5) (1997) 375-381

[123] C.C. Leffler, B.UW Müller, Chitosan-gelatin sponges for controlled drug delivery: the use of ionic and non-ionic plasticizers. S.T.P. Pharma Sciences 10 (2000) 105-111

[124] S-B. Park, J-O. You, H-Y. Park, S.J. Haam, W-S. Kim, A novel pH-sensitive membrane from chitosan-TEOS IPN: preparation and its drug permeation characteristics. Biomaterials 22 (2001) 323-330

[125] Y. Nishioka, S. Kyotani, M. Okamura, Y. Mori, M. Miyazaki, K. Okazaki, S. Ohnishi, Y. Yamamoto, K. Ito, Preparation and evaluation of albumin microspheres and microcapsules containing cisplatin. Chemical and Pharmaceutical Bulletin 37(5) (1989) 1399-1400; Y. Nishioka, S. Kyotani, H. Masui, M. Okamura, M. Miyazaki, K. Okazaki, S. Ohnishi, Y. Yamamoto, K. Ito, Preparation and release characteristics of cisplatin albumin microspheres containing chitin and treated with chitosan. Chemical and Pharmaceutical Bulletin 37(11) (1989) 3074-3077; Y. Nishioka, S. Kyotani, M. Okamura, M. Miyazaki, K. Okazaki, S. Ohnishi, Y. Yamamoto, K. Ito, Release characteristics of cisplatin chitosan microspheres and effect of containing chitin. Chemical and Pharmaceutical Bulletin 38(10) (1990) 2871-2873

[126] M. Açikgöz, H.S. Kaş, Z. Hasçelik, Ü. Milli, A.A. Hincal, Chitosan microspheres of diclofenac sodium, II: In vitro and in vivo evaluation. Pharmazie 50 (1995) 275-277

[127] K.C. Gupta, M.N.V. Ravi Kumar, Drug release behavior of beads and microgranules of chitosan. Biomaterials 21 (2000) 1115-1119

42

128   J. Akbuğa, N. Bergişadi, 5-Fluorouracil-loaded chitosan microspheres: preparation and release characteristics. J. Microencapsulation 13(2) (1996) 161-168; S. Patashnik, L. Rabinovich, G. Golomb, Preparation and evaluation of chitosan microspheres containing bisphosphonates. J. Drug Targeting 4(6) (1997) 371-380; H. Tokimitsu, H. Ichikawa, T.K. Saha, Y. Fukumori, L.H. Block, Design and preparation of gadolinium-loaded chitosan particles for cancer neutron capture therapy. S.T.P. Pharma Sciences 10 (2000) 39-49

129   I.M. van der Lubben, J.C. Verhoef, A.C. can Aelst, G. Borchard, H.E. Junginger, Chitosan microparticles for oral vaccination: preparation, characterization and preliminary *in vivo* uptake studies in murine Peyer's patches. Biomaterials 22 (2001) 687-694

130   L. Illum, P. Watts, A.N. Fischer, I. Jabba Gill, S.S. Davis, Novel chitosan-based delivery systems for the nasal administration of a LHRH-analog. S.T.P. Pharma Sciences 10 (2000) 89-94

131   K.W. Leong, H-Q. mao, V.L. Troung-Le, K. Roy, S.M. Walsh, J.T. August, DNA-polycation nanospheres as non-viral gene delivery vehicles. J. Controlled Release 53 (1998) 183-193

132   J. Murata, Y. Ohya, T. Ouchi, Possibility of application of quaternary chitosan having pendant galactose residues as a gene delivery system. Carbohydrate Polymers 29 (1996) 69-74

133   J. Murata, Y. Ohya, T. Ouchi, Design of quaternary chitosan conjugate having antennary galactose residues as a gene delivery tool. Carbohydrate Polymers 32 (1997) 105-109

134   K.Y. Lee, I.C. Kwon, Y-H. Kim, W.H. Jo, S.Y. Jeong, Preparation of chitosan self aggregates as a gene delivery tool. J. Controlled Release 51 (1998) 213-220

135   F-L. Mi, C-T Chen, Y-C Tseng, C-Y. Kuan, S-S Shyu, Iron(III)-carboxymethylchitin microsphere for the pH-sensitive release of 6-mercaptopurine. J. Controlled Release 44 (1997) 19-32

136   P. Giunchedi, I. Genta, B. Conti, R.A.A. Muzzarelli, U. Conte, Preparation and characterization of ampicillin loaded methylpyrrolidinone chitosan and chitosan microspheres. Biomaterials 19 (1998) 157-161

137   B. Conti, P. Giunchedi, I. Genta, U. Conte, The preparation and *in vivo* evaluation of the wound-healing properties of chitosan microspheres. S.T.P. Pharma Sciences 10 (2000) 101-104

138   C. McQueen, A. Silvia, P.K. Lai, M. Amiji, Surface and blood interaction properties of poly(ethylene oxide)-modified chitosan microspheres. S.T.P. Pharma Sciences 10 (2000) 95-100

139   R.A.A. Muzzarelli, M. Miliani, M. Cartolari, I. Genta, P. Perugini, T. Modena, F. Pavanetto, B. Conti, Oxychitin-chitosan microcapsules for pharmaceutical use. S.T.P. Pharma Sciences 10 (2000) 51-56

140   Y. Ohya, R. Cai, H. Nishizawa, K. Hara, T. Ouchi, Preparation of PEG-grafted chitosan nanoparticles as peptide drug carriers. S.T.P. Pharma Sciences 10 (2000) 77-82

141   F.N. Bruscato, A.G. Danti, Pharmaceutical tablets containing chitin as a disintegrant", US Patent 4086335, April 25th, 1978

142   A.B. Thakur, N.B. Jain, Controlled release formulation and method US Patent 4738850, April 19th, 1988

143   F. Sekigawa, Y. Onda, Coated solid medicament form having releasability in large intestine, US Patent 5217720, June 8th, 1993

144   T. Suzuki, K. Hashiudo, T. Matsumoto, T Higashide, Large intestinal dissociative hard capsules, US Patent 5468503, February 1st. 1994; A. Yamada, T. Wato, N. Uchida, M. Fujisawa, S. Takama, Y. Inamoto, Oral pharmaceutical preparation released at infragastrointestinal tract, US Patent 5468503, November 21st. 1995

145   M.M. Daly, R.W. Keown, D.W. Knorr, Chitosan alginate capsules, US Patent 4808707, February 28th, 1989

146   S. Magdassi, K. Mumcuoglu, U. Bach, Method of preparing natural-oil-containing emulsions and microcapsules and its uses, US Patent 5518736, May 21st, 1996; S. Magdassi, K. Mumcuoglu, U. Bach, Y. Rosen, Method of making positively charged microcapsules of emulsions of oils and its uses, US Patent 5753264, May 19th, 1998

147   P. J. Watts, L. Illum, Drug delivery composition containing chitosan or derivative thereof having a defined Z-potential, US Patent 5840341, November 24th, 1998

148   M. Hashimoto, M. Otagiri, T. Imai, Drug composition, US Patent 5474989, December 12th, 1995

149   F. Conti, Chitosan derivatives in the form of coordinated complexes with ferrous ions, US Patent 4810695, March 7th, 1989

150   L. H. Block, S. S. Sabnis, Methods of creating a unique chitosan and employing the same to form complexes with drugs, delivery of the same within a patient and a related dosage form, US Patent 5830883, November 3rd, 1998; L. H. Block, S. S. Sabnis, Methods of creating a unique chitosan and employing the same to form complexes with drugs, delivery of the same within a patient and a related dosage form, US Patent 5900408, May 4th 1999

151   I. Azuma, S. Tokura, S. Nishimura, H. Seo, Slow-releasing preparation, US Patent 4873092, October 10th, 1989

152   K. Kifune, Method for manufacture of biodegradable drug donor and drug donor made thereby, US Patent 4704268, November 3rd, 1987

153   K. Kifune, K. Motosugi, H. Tanae, Method for preparation of a shaped chitin body containing a physiologically active substance, US Patent 5290752, March 1st, 1994

154   C. P. Chung, S. J. Lee, Biodegradable sustained release preparation for treating periodontitis, US Patent 5855904, January 5th, 1999

155   J. R. Cardinal, W. J. Curatolo, C. D. Ebert, Chitosan compositions for controlled and prolonged release of macromolecules, US Patent 4895724, January 23rd, 1990

156   K. Kifune, K. Motosugi, H. Tanae, Method for preparation of a shaped chitin body containing a physiologically active substance, US Patent 5290752, March 1st, 1994

157   J. M. Gallo, E. E. Hassan, Receptor-mediated delivery system, US Patent 5129877, July 14th, 1992

158   L. Illum, Systemic drug delivery compositions Ccmprising a polycationic substance, US Patent 5554388, September 10th, 1996; L. Illum, Drug delivery composition, US Patent 5744166, April 28th, 1998; L. Illum, Composition for nasal administration, US Patent 5629011, May 13th, 1997

159   S. N. Chatfield, Vaccine compositions, US Patent 6048536, April 11th, 2000

160   M. M. Amiji, Drug delivery using pH-sensitive semi-interpenetrating network hydrogels, US Patent 5904927, May 18th, 1999

161   Information retrieved from website: www.devicelink.com in March 2001

162   Information retrieved from website: www.bccresearch.com in March 2001

163   Bandage made from algae can help stop bleeding. The Straits Times, May 9, 2001, p10.

44

[164]   5th ICCC Round Table Discussion. in Advances in Chitin and Chitosan, C.J. Brine, P.A. Sandford, J.P. Zikakis, eds., Elsevier Applied Science, New York, N.Y., 1992. 671

CHAPTER 3: CHITIN AS A BIOMATERIAL

3.1 BIOMATERIALS

An important criterion in the success of any medical device is the material used to make these devices as the material interact directly with the tissue. The term biomaterial has come to encompass all materials used in these situations and has been defined as follows.

"A biomaterial is a nonviable material used in a medical device, intended to interact with biological systems".[1]

In its broadest sense, a biomaterial is a material used for biomedical applications i.e. in a clinical situation. Metals, ceramics and polymers have all been applied as biomaterials, with polymers making progressive impact since the end of the Second World War. Polymers are of two types, synthetic and natural. The use of synthetic polymers in biomedical applications have grown in popularity as the situations when and where they can be applied have generally been defined leading to eventual widespread acceptance over the past 40-50 years. Normally, synthetic polymers are non-degradable, but are subject to wear. In instances where they are degradable, such as poly-lactic acid, poly-glycolic acid and their co-polymers in degradable sutures, their degradation mechanisms and degradation products are quite well established.

Biopolymers on the other hand have a more recent history with cellulose being the exception, having been used as a gauze material in dressings for decades. An important reason for using biopolymers lies in the rationale that materials derived from nature will exhibit greater compatibility with humans. This is an apparent advantage not synonymous with synthetic polymers. With this in mind, biopolymers as constituents in animal tissues have been used as bioprostheses, motivated by the need to find replacements that match as closely as possible to humans to hasten the restoration of function with minimal complications. For example, processed pericardial tissue and the pig aortic heart valve have been used since the late 1960's in bioprosthetic (or tissue) heart valves, cardiac patches and vascular grafts.[2] The tissue heart valve, essentially a collagen-based assembly derived from the pig aortic heart valve performs better than its mechanical counterpart with respect to blood compatibility attributed to its natural shape. However, regulatory requirements have also become more rigorous with concerns over livestock diseases, and biopolymers derived from animal sources are brought under stringent scrutiny. Despite these concerns, biopolymers such as collagen and hyaluronic acid have been introduced for a variety of applications in recent years, progressing from the use of whole tissue in bioprosthesis to the utilization of biopolymers extracted from tissue. The advantage of the biopolymer form are high purity and processing utility. Furthermore, in situations where biodegradation is warranted, the pathways for biodegradation can be more predictable when in pure form.

3.2 CHITIN'S ROLE AS A BIOMATERIAL

Onto this stage, proponents have propelled chitin with lofty expectations. The potential for chitin to play a biomedical role has been outlined and indisputably the prospect is great. Chitin as a biomaterial can be exploited in 2 main manners, as bio-stable chitin or as a biodegradable material. The five biomedical applications surveyed for chitin are neatly divided as follows. Wound dressings and coatings onto blood contact tubing are external device situations where chitin is used in a "bio-stable" form. Bone substitutes and tissue engineering applications are classified in the implant category. These applications require chitin to participate in a biodegradation role, the purpose being to avoid the necessity for a second operation to retrieve the implant. A drug delivery role also implies that retrieval is not usually an option.

3.2.1 Bio-Stable Chitin

Chitin and chitosan are not normally water soluble (with the exception of low molecular weight components) at the body's pH of 7 but can be subject to effects of erosion caused by constant interaction with bodily fluids for example. However, chitin is susceptible to biodegradation by the body's enzymes or defensive systems.

For a bio-stable role, chitin will be expected to maintain its integrity throughout the period of use. The most credible method to make chitin non-degradable is by chemical derivatization although strictly speaking, a material will degrade when used for an extended term. Chemical derivatization gives rise to structures not readily recognizable by enzymes and therefore the chitin is more degradation resistant.

The wound dressing applications for chitin is designed in most instances to be an external device that does not biodegrade. While it is claimed that lysozyme abundantly present in wound beds will degrade chitin wound dressings, this may only happen at the interface, leaving the dressing mostly intact. When used as a coating for blood-contact tubing, the substituted chitins and chitosans are expected to remain on the inner surface of the lumen, provided of course, proper anchoring of the chitin to that surface was achieved.

3.2.2 Biodegradable-Chitin

Implicit in the concept of a biodegradable biomaterial is the ability for the biomaterial to perform its required function for a predetermined time period with the gradual dissipation of the biomaterial until ultimately, it is totally assimilated or disposed by the body. The term **biodegradable** normally refers to a material being susceptible and degraded by enzymes and other bio-based reactions when placed in the biological system. The term further implies that the deterioration of the material is controlled at a rate that is desirable for the material to perform its biomaterials role. However, biodegradable is not necessarily equivalent to **bioerodible**. Bioerodible typically describes the situation where the material is being hydrolyzed or dissolved by aqueous media in the biological environment, suggesting physical erosion rather than biochemical action. Chitin-based implants will be exposed to both mechanisms *in vivo*, with biodegradation, presumably dominating.

The biodegradability of chitin into non-harmful residues has been a popular platform for projecting the biopolymer as a biomaterial extraordinaire. What is the basis for this optimism? It is fitting to begin by reviewing the biological origin of chitin and how this impacts on its biodegradation property.

3.2.2.1 Biosynthesis and Biodegradation of Chitin in Nature

The estimated amount of chitin and chitosan synthesized by all known biological systems that produce these biopolymers is in the region of one billion tons per year.[3] It is self-evident that in nature, chitin is synthesized and subsequently recycled, both processes mediated by enzymes. Chitin biosynthesis begins with glucose that is converted by several enzymatic reaction steps into glucosamine-6-P sugar that is subsequently N-acetylated, and finally forms uridine diphosphate-N-Acetylglucosamine (UDP-GlcNAc) the precursor of chitin, as depicted in Figure 3.1.[4]

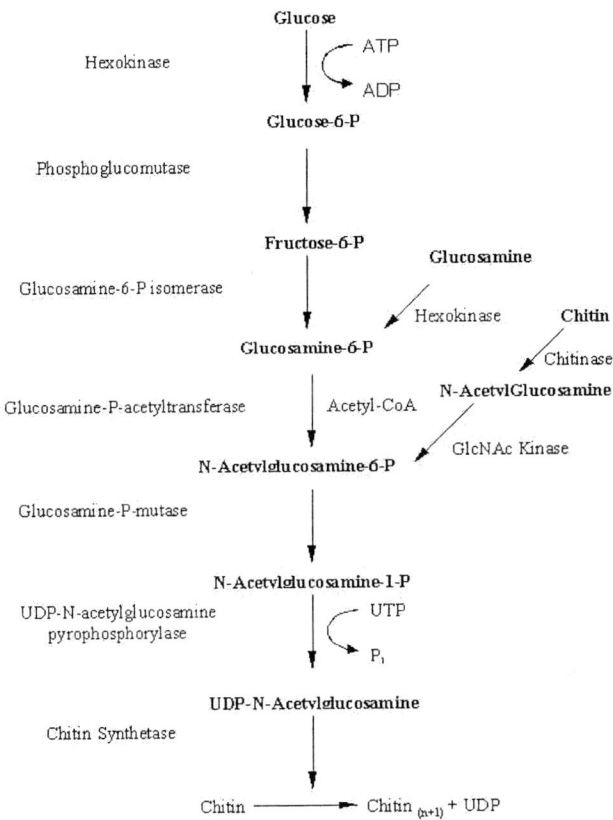

Figure 3.1:   Generalize metabolic pathway for the synthesis of UDP-GlcNAc
(Reprinted with permission from CRC Press)

At this juncture, the principal enzyme is chitin synthetase that will add GlcNAc to chitin oligosaccharides or primer molecules.[5]   How this is achieved depends on whether the organism is animal or plant as the organization of chitin in crustaceans is different from fungi.   In crustaceans, invariably there is some association with proteins via chemical bonding, a result of the intimate relationship of the structural role both chitin and protein play in the cuticle or shell of the animal.   In fungi, the formation of chitin requires the consideration of how the growing chitin chain organizes itself via hydrogen bonding to be deposited as microfibrils in the cell wall.[6] Finally in the instance of chitosan containing fungi, chitin deacetylase converts chitin into chitosan.[7]

Chitin is a very resilient biopolymer that is very difficult to breakdown. It is even resistant to degradation by mammalian digestive enzymes.[8] Nevertheless, there is no accumulation of chitin in nature, invoking the question how this is managed by nature. The answer was first found in the discovery of bacterial enzymes, chitinases that can breakdown chitin to maintain the natural balance in nature.[9] Chitinases are now known to be present in the digestive tract of many invertebrates and vertebrates including fishes, lizards, birds and mammals such as the guinea pig.[10] Chitinases have also been found in higher plants where cellulose, and not chitin, forms the structural component. This occurrence in plants may be a self-defense mechanism against pathogenic microbes and insects that have chitin as a component in their cell wall or exoskeleton. In humans, the role of chitinase appears to be delegated in part to the enzyme, lysozyme.[11] Last, chitosanase enzymes effect the biodegradation of chitosan in nature.

The general biodegradation pathway of chitin in nature is summarized in Figure 3.2. The first step involves the breakdown of the long chitin chain by chitinases into smaller oligomers such as diacetylchitobiose. Subsequently, more specific enzymes such as β-acetyl-N-hexosaminidases or chitobioses hydrolyze the oligomers to *N*-acetyl-D-glucosamine that is readily utilized by microorganisms found in the flora and fauna. Similarly, chitosanases first

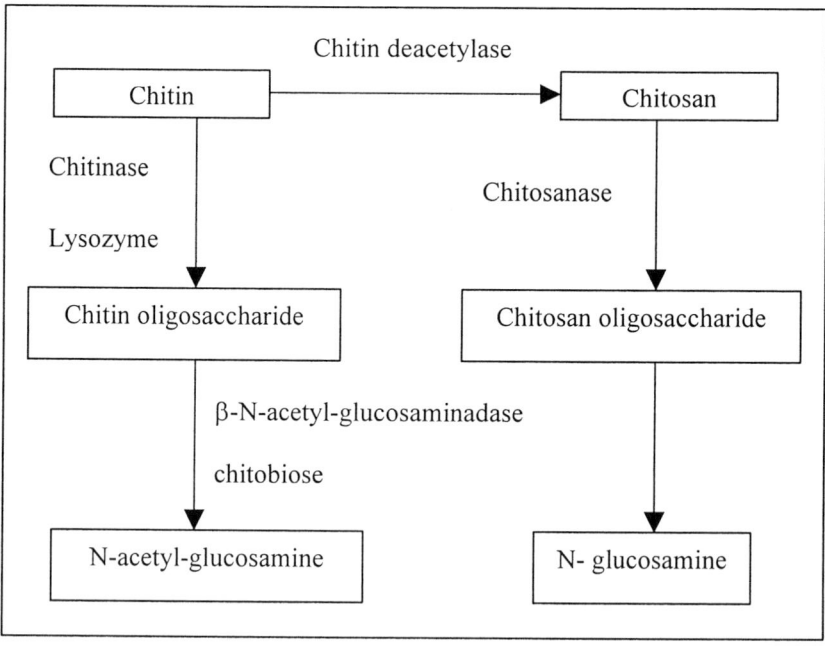

Figure 3.2:   Biodegradation of chitin in nature

break chitosan down into oligomers that are further acted upon by other enzymes to generate N-glucosamine. In essence, nature has implemented a system of chitin activity that preserves the eco-balance of chitin and nitrogen/carbon utilization. Therefore, as far as nature is concerned, chitin is both biodegradable and environmentally friendly.

From the foregoing synopsis, two questions come to mind. Does this picture change upon utilizing chitin outside the eco-system? If so, what affect does it have on chitin as a biodegradable biomaterial?

The answer to the first question is YES. First as has been noted, the seafood processing industry generated a large quantity of waste from the shellfish industry. This brought the eco-control system into unbalance, as it was not designed to manage waste of that scale in a limited timeframe. One outcome in addressing the eco-imbalance is the advent of chitin. Second, in the process of isolating chitin from shellfish, large quantities of hydrochloric acid and sodium hydroxide were used and resulted in a new pollution problem. This has led to measures in wastewater treatment and cleaner alternatives such as enzyme and microbial fermentation. Despite the challenges, it is the author's opinion that environmental friendliness of chitin outside the eco-system will prevail eventually.

To understand the impact the biodegradation property has on chitin as a biodegradable biomaterial, we take a look at what is the likely scenario for chitin when implanted in humans.

3.2.2.2 Biodegradation of Chitin in Humans

From the human body viewpoint chitin is not native, although it's constituents N-acetylglucosamine and N-glucosamine are.[12] However, there are several enzyme systems present in humans that can break chitin down. First, macrophages in the body do contain chitinases and lysozyme as part of their phagocytic arsenal.[13] In addition, the terminal non-reducing N-acetylglucosamine residue of chitin is also exposed to hydrolysis by N-acetyl-β-D-glucosaminidase. Finally, chitosans with varying degrees of deacetylation are susceptible to several human hydrolases. Studies employing some of these enzyme systems abound supporting the susceptibility of chitin to biodegradation *in vivo* and have been used to advocate that chitin is biodegradable in humans.

The most numerous reports on the *in vitro* biodegradation of chitin with reference to humans have centered on the enzyme lysozyme with the majority of studies using hen egg-white lysozyme as a convenient source of the enzyme.[14] One of the first studies was the hydrolysis of highly purified chitin by lysozyme into N-acetyl-glucosamine. The concentration of N-acetylglucosamine was observed to increase linearly with time for up to 30 hours. In the same study, bacterial chitinase was shown to exert an approximately similar effect. On the other hand, neither enzyme degraded chitosan.

The specificity of lysozyme for high molecular weight chitin was elucidated from a study of the effect of several lysozyme types on chitopentaose.[15] This specificity of lysozyme action was noted by the lysozymic resilience of synthetically derived chitin or acetamidodeoyxcellulose that had a low concentration of N-acetyl groups.[16] Amano and Ito noted a high degree of N-acetylation on chitin as necessary for the effective action of lysozyme in chitin biodegradation. Their results showed that a poor degree of N-acetyl and amino groups at the C-2 position led to non-detectable hydrolysis of the synthetic chitin.[17]

Hirano et al added to this conclusion with their report that partially N-acetylated chitosan of degree of substitution 0.2, 0.4, 0.6 and 0.8 were up to 15 times more readily hydrolyzed by lysozyme than fully acetylated chitin.[18] Sashiwa et al formed similar conclusions with their study of partially deacetylated chitins'.[19] Aiba also showed that moderately (20-30%) N-deacetylated chitins (MDC) were more hydrolysable than partially (20-70%) N-acetylated chitosans (HAC).[20] Aiba attributed this to the distribution of N-acetyl groups in the chain, MDC having block type GlcNAc and D-glucosamine distribution while the HAC had a more random arrangement of the GlcNAc and D-glucosamine residues. The blocks of GlcNAc (3 to 5 in a row) were necessary for efficient lysozyme action.

The crystallinity of chitin also plays a role. Shigemasa et al found β-chitin to be more readily digested compared to α-chitin.[21] The same group also noted that chitin deacetylated under homogeneous conditions were more susceptible to lysozyme compared to those deacetylated under heterogeneous conditions. Many crystalline regions are left intact during heterogeneous deacetylation, and not accessible by enzymes. Kurita et al have recently reported on the ready susceptibility of β-chitin to lysozyme and related this ease of degradation to the degree of deacetylation. The rate of degradation was found to be a maximum at 50% degree of deacetylation.[22] Interestingly, the biodegradation of chitin by lysozyme is enhanced by pretreatment with organic chemicals.[23]

Finally, two recent studies highlight the ready biodegradability of chitin *in vivo* using animal model studies. In the report by Saimoto et al, chitin from squid pen were very sparingly digested by lysozyme, yet were completely absorbed in 14 days when implanted subcutaneously in dogs.[24] The lack of degradation in lysozyme was perplexing to the authors who suggested that other agents in the animal model were responsible for degrading the 90% acetylated squid pen chitin. Onishi and Machida reported that 50% deacetylated chitin delivered by intraperitoneal injection into mice was excreted as small molecular weight materials in the urine.[25] The results clearly indicate that chitin is degraded *in vivo* and does not accumulate in the body.

In studies using chitinase derived from bacterial sources, chitinase activity is not necessarily triggered by the N-acetyl sequence in the chitin chain.[26] There is now increasing evidence that human chitinases are present. Escott and Adams have reported chitinase type activity in human serum and leukocytes.[27] The chitinolytic activity was demonstrated to be distinct from that of lysozyme.

Vårum et al showed that human serum degrades water-soluble chitosan, the extent increasing with the degree of N-acetylation.[28] The depolymerization was attributed to lysozymes present in the human serum.

Ikada et al have reported one of the most comprehensive studies of the biodegradation of chitin and chitosan.[29] Chitins of varying degrees of deacetylation were subjected to lysozyme *in vitro* degradation and *in vivo* rat model evaluation. Samples with a degree of deacetylation below ~70% were readily degraded by lysozyme, in the animal and histology showed acute inflammation consistent with rapid biodegradation. Samples with a degree of deacetylation above ~70%, were poorly degraded with chitosan totally not degraded. This differences in enzyme susceptibility bordering the 70% degree of deacetylation mark was attributed to the sequential arrangement of the N-acetylglucosamine monomer unit that is necessary for identification by the enzymes and again allude to chitosan as not being degradable *in vivo*.

From the foregoing, it is evident that chitin is susceptible to enzymatic degradation. Chitosan's susceptibility to lysozyme is dependent on its degree of N-acetylation. The presence of lysozyme and other chitin active enzymes, especially chitinase and other hydrolases that can participate in biodegradation *in vivo*, in the human body has also been established and is reassuring. In fact, most materials can be phagocytized by macrophages in response to foreign bodies.[30] In the author's observation, a term "bio-attacked" appears more appropriate to describe chitin-based materials *in vivo* as inferred from a series of intramuscular biocompatibility studies of chitin implants where the implants deteriorated in stages into oblivion in a process that can hardly be straightforward as to classify into biodegradation and bioerosion![31] The connotation of these observations augurs well for chitin as a biodegradable biomaterial.

3.2.2.3 Implications of *In Vitro* Enzyme Biodegradation Studies on Chitin to Humans and Biomedical Applications

The extrapolation of *in vitro* enzyme biodegradation and *in vivo* small animal model studies, to humans must be circumspectly approached. Why?

First, while the preceding discussions do form a collection of good data that is indicative of chitin biodegrading in humans, the chitin samples used in these studies were derived from a collage of commercial grades subsequently purified or were extracted in individual laboratories. The non-uniform nature of the chitins used in these experiments suggests that comparisons are at best qualitative and definitive conclusions cannot be formulated, warranting more structured and controlled studies as follow up. Essentially, a necessity exists for properly characterized starting materials with defined properties such as molecular weight and degree of acetylation, be it made available from producers or in-house derived, prior to investigations especially in biodegradation studies. This is requisite to convincingly establish the biodegradable character and also the degradation pathways for chitin.

Second, the very inhomogeneous character of chitin, its variety of possible chemical forms and the wide range of physical characteristics such as microstructure, molecular weight and degree of acetylation/deacetylation, all contribute a myriad of factors that can influence their biodegradation by all enzyme types. Whether a particular chemical derivative is even susceptible to biodegradation is suspect and for a bio-stable chitin role, should be demonstrated as "non-degradable". The complexity of degradation products of these forms must be evaluated. This will be useful with the increasing sophistication in chitin derivatization methods where new and better chitin-based biomaterials may be in the offing that are designed to biodegrade and have good interactions with other materials such as pharmaceuticals or cells. The design of proper protocols for both *in vitro* and *in vivo* studies should be established and performed with a database possibly created, cataloging each individual form's biodegradability and biodegradation products.

Finally, all known studies have focused on chitin and chitosan with varying degrees of acetylation using individual enzymes, primarily lysozyme and chitinase. This focus may be restrictive, because focus on lysozyme precludes the opportunity to search for other enzymes for chitin biodegradation in humans to complete the story on biodegradability. In other words, the study of the biodegradation of chitin in humans should be all-inclusive, and not limited to one enzyme system as it is too idealized. As the animal model work has shown, chitin is resorbed *in vivo* although the *in vitro* studies are to some extent, inconclusive. This may expose biomedical applications of chitin to complications that so far have not been addressed. Perhaps new studies could involve enzymatic cocktails in differing permutations to appraise this effect on biodegradability.

3.2.2.4 Biodegradation Lifetime and Fate of Biodegradation Products

In nature, biodegradation is time independent as long as the eco-system is kept in balance. In biomedical applications, the time scale of biodegradation is critical. For example when chitin is used as a drug delivery carrier, an important question arises of whether drug release is dependent on biodegradation. This is because biodegradation in a controlled manner and at a constant rate is difficult to achieve. These issues must be considered thoroughly if the material is to satisfy the role of drug delivery carrier. In addition, the residues of biodegradation must be harmless, and this cannot be generalized, especially to all chitin derivatives. Therefore, in order for biodegradation to be meaningful in biomedical applications, each and every chitin-based material must be properly characterized to ensure its safe use on humans.

3.3 INFERENCE

Chitin as a biomaterial is plausible, both as bio-stable chitin and biodegradable chitin. The less intuitively obvious aspect when exploiting chitin as a biomaterial is that chitin must assume many varied physical and chemical forms with its attendant challenges. This is both providential as well as a portentous. Providential as the challenges in resolving all these issues ensures that chitin science and technology will grow. Portentous as the magnitude of the task to establish chitin as a biomaterial is massive. Apart from the obvious technical aspects of manipulating chitin into various usable forms, biomedical-based issues key of which is biocompatibility and non-toxicity must be addressed.

3.4 REFERENCES

[1]   D.F. Williams, Definitions in Biomaterials. Proceedings of a Consensus Conference of the European Society for Biomaterials, Chester, England, March 3-5 1986 Vol.4 Elsevier, New York, 1987.

[2]   E. Khor, Methods in the treatment of collagenous tissues for bioprostheses. Biomaterials, 18 (1997) 95-105

[3]   S. Hirano, H. Inui, H. Kosaki, Y. Uno, T. Toda, Chitin and chitosan: Ecologically bioactive polymers. in Biotechnology and Bioactive Polymers, C. Gebelein, C. Carraher, eds., Plenum Press, New York, USA, 1994. 43-54

[4]   M.N. Horst, A.N. Walker, E. Klar, The pathway of crustacean chitin synthesis. in The crustacean integument: Morphology and biochemistry, M.N. Horst, J.A. Freeman, eds., CRC, Boca Raton, FL, USA, 1993. 113-149

[5]   C. Jeuniaux, M.F. Voss-Foucart, Chitin Biomass and production in the marine environment. Biochemical Systematics and Ecology, 19 (1991) 347-356

[6]   J. Ruiz-Herrera, Chitin and chitosan. in Fungal cell wall: structure, synthesis, and assembly, CRC Press, Boca Raton, FL, USA, 1992. 89-117

[7]   E. Cabib, The synthesis and degradation of chitin. in Advances in Enzymology and Related Areas of Molecular Biology, Vol. 39, A. Meister, ed., John Wiley & Sons, New York, N.Y., 1987. 59-101

[8]   A.G. Richards, The decomposition of chitin and cuticle in nature. in The integuments of arthropods: The chemical components and their properties, the anatomy and

development, and the permeability, University of Minnesota Press, Minneapolis, MN, 1951. 54

[9] M.V. Deshpande, Enzymatic degradation of chitin and its biological applications. J. Science and Industrial Research 45 (1986) 273-281

[10] C. Jeuniaux, An addition to the list of hydrolases in the digestive tract of vertebrates. Nature 192 (1961) 135-136

[11] G.W. Gooday, Aggressive and defensive roles for chitinases, in Chitin and chitinases, P. Jollès, R.A.A. Muzzarelli, eds., Birkhaüser-Verlag, Basel, Switzerland, 1999. 157-169

[12] R.A.A. Muzzarelli, Human enzymatic activities related to the therapeutic administration of chitin derivatives, Cellular and Molecular Life Sciences, 53 (1997) 131-140

[13] R.G. Boot, G.H. Renkema, A. Strijland, A.J. van Zonneveld, J.M.F.G. Aerts, Cloning of a cDNA encoding chitotriosidase, a human chitinase produced by macrophages, J. Biological Chemistry 44 (1995) 26252-26256

[14] L.R. Berger, R.S. Weiser, The β-glucosaminidase activity of egg-white lysozyme. Biochimica et Biophysica Acta 26 (1957) 517-521

[15] D. Charlemagne, P. Jollès, The action of various lysozymes on chitopentaose. FEBS Letters 23(2) (1972) 275-278

[16] S. Hirano, Y. Kondo, K. Nagamura, Acetamidodeoxycellulose and its digestibility by chitinase and lysozyme. International J. Biological Macromolecules 9 (1987) 308-310

[17] K-i. Amano, E. Ito, The action of lysozyme on partially deacetylated chitin. European J. Biochemistry 85 (1978) 97-104

[18] S. Hirano, H. Tsuchida, N. Nagao, N-acetylation in chitosan and the rate of its enzymatic hydrolysis. Biomaterials 10 (1989) 574-576

[19] H. Sashiwa, H. Saimoto, Y. Shigemasa, R. Ogawa, S. Tokura, Lysozyme susceptibility of partially deacetylated chitin. International J. Biological Macromolecules 12 (1990) 295-296

[20] S-i. Aiba, Studies on chitosan: 4. Lysozymic hydrolysis of partially N-acetylated chitosans. International J. of Biological Macromolecules 14 (1992) 225-228

[21] Y. Shigemasa, K. Saito, H. Sashiwa, H. Saimoto, Enzymatic degradation of chitins and partially deacetylated chitins. International J. Biological Macromolecules 16 (1994) 43-49

[22] K. Kuria, Y. Kaji, T. Mori, Y. Nishimiya, Enzymatic degradation of β-chitin: susceptibility and the influence of deacetylation. Carbohydrate Polymers 42 (2000) 19-21

[23] T. Morita, H.J. Lim, I. Karube, Enzymatic hydrolysis of polysaccharides in water-immiscible organic solvent, biphasic systems. J. Biotechnology 38 (1995) 253-261

[24] H. Saimoto, Y. Takamori, M. Morimoto, H. Sashiwa, Y. Okamoto, S. Minami, A. Matsuhashi, Y. Shigemasa, Biodegradation of chitin with enzymes and vital components. Macromolecular Symposium 120 (1997) 11-18

[25] H. Onishi, Y. Machida, Biodegradation and distribution of water-soluble chitosan in mice. Biomaterials 20 (1999) 175-182

[26] S-i. Aiba, Studies on chitosan: 6. Relationship between N-acetyl group distribution pattern and chitinase digestibility of partially N-acetylated chitosans. International J. Biological Macromolecules 15 (1993) 241-245

[27] G.M. Escott, D.J. Adams, Chitinase activity in human serum and lecukocytes. Infection and immunity 63(12) (1995) 4770-4773

28   K.M. Vårum, M.M. Myhr, R.J.N. Hjerde, O. Smidsrød, *In vitro* degradation rates of partially N-acetylated chitosans in human serum. Carbohydrate Research 299 (1997) 99-101

29   K. Tomihata, Y. Ikada, *In vitro* and *in vivo* degradation of films of chitin and its deacetylated derivatives. Biomaterials 18 (1997) 567-575.

30   C.S. Kim, L.J. Folinsbee, Physiological and biomechanical factors relevant to inhaled drug delivery. in Lung Biology in Health and Disease. Vol 107: Inhalation Delivery of Therapeutic peptides and Proteins, A.L. Adjei, P.K. Gupta, eds., Marcel Dekker Inc., New York, N.Y., 1997. 3-25

31   A-S. Baugenard, L.Y. Lim, A. Wee, E. Khor, Biocompatibility evaluation of hydroxyapatite-chitin materials. 8th International chitin and chitosan conference, 2000. Abstract: B-5, 60

CHAPTER 4: BIOCOMPATIBILITY ISSUES

4.1 WHAT IS BIOCOMPATIBILITY?

For a biomaterial to be approved for use in contact with the human body, especially as a component of a medical device such as an implant to replace a defective or diseased tissue and in drug delivery applications, the acceptance of the biomaterial by the human biological system has to be demonstrated. These events occur every time a new device and its associated biomaterial(s) are proposed for use and also on a routine basis as a quality control safeguard throughout the product life cycle. The authority to grant approval to accept the medical device and its associated biomaterial(s) used in fabricating the device lie with the Health Regulating body in the market where the end-use device is sold. Each market's requirements can be different, although the present trend towards international harmonization of Standards may eventually result in multiple markets accepting the same requirements. The bottom line is, no biomaterial can be considered for use in contact with human tissue unless it has been proven acceptable.

What constitutes this acceptability? For a medical device and its component biomaterials to be acceptable, the producer has to demonstrate that the product is safe and will perform the role as intended. For chitin, the basic premise is as a biomaterial that will be further processed into a final form. This compels the supplier of chitin raw material and the manufacturer of the chitin containing medical device or product to conform to regulatory requirements. What is the premise for determining acceptability? Safety and performance of a medical device are paramount, generally encompassed by the term biocompatibility.

Biocompatibility is defined as "the ability of a biomaterial to perform with an appropriate host response in a specific application".[1] Biocompatibility is concerned with the interactions that occur between biomaterials and host tissue. The effect of the biomaterial on tissues and vice versa has to be understood to ensure safety and performance of the device. Safety is primarily concerned with toxicity issues including degradation products while performance deals mainly with satisfying the role the medical device was designed for. Biocompatibility as defined is situation dependent and should not be used to imply that the biomaterial is acceptable in general terms i.e. all biomedical applications, as a material can fail when applied differently. In other words, "no material is unequivocally biocompatible".[1] For a medical device and its associated biomaterial(s), the process in which biocompatibility is ascertained lies in establishing reliable, reproducible and verifiable methods of evaluation that include biological, chemical, mechanical and physical tests.

The evaluation of biocompatibility has been defined, developed and refined over the past 30 or so years and is now well entrenched. The important issues are centered on the host reactions to biomaterials and vice versa as summarized in Table 4.1. The demonstration of biocompatibility includes safety focused studies as in exposing the biomaterial to a series of both in vitro and in vivo studies to establish that it is non-cytotoxic and compatible with the specific functions of the cell and animal models. Other important considerations include the stability of the biomaterial over the anticipated residence time the device is used and the statistical size required for the data to be meaningful due to the experimental variability. Examples of performance studies are studying the strength of the material over the anticipated period of use and verifying its mechanical functions in a simulated body fluid environment for a mechanical device.

Effect ON host by biomaterial	Effect BY host on biomaterial
Local	Biological
Toxicity	Physical
Healing	Mechanical
Systemic and remote	

Table 4.1:   Some common criteria for assessing biocompatibility

A survey of the scientific literature suggests that the term biocompatibility has often been generalized. That is to say, very frequently, performance or function studies of only one or two aspects of a crude model representing a medical device sometimes intermingled with one or two safety evaluations are reported. These studies are good indicators but are not necessarily a comprehensive representation of the complete biocompatibility profile as defined above. Therefore the term biocompatibility has to be used expediently so as to avoid attaching an overwhelming value that may not stand up to careful scrutiny. This is especially easy to do with a biomaterial such as chitin where the biocompatibility term has been freely associated with the biopolymer and its derivatives. How did this impression originate and is it appropriate?

4.2 IS CHITIN REALLY BIOCOMPATIBLE?

Chitin and chitosan have indeed captured the imagination of the scientific community, the producers of chitin and chitosan, and manufacturers of chitin-based materials and products for biomedical applications. Over the years, the close relationship with its biological, hence natural origins, have led to the labeling of chitin and chitosan as being environmentally friendly, biodegradable and biocompatible. As the number of papers, exchanges and references to the biopolymers grew, this close connection of chitin with the labels has somewhat self-perpetuated into respected acceptance. Consequently, chitin has become *de facto* biocompatible and advocated for biomedical applications. After all, many people eat shrimp shells and survive! The matter at hand is whether the association of chitin and chitosan with the term biocompatibility comes from pure generalization or concluded from test results that meet the true requirements of biocompatibility.

To answer the origin question of chitin's *de facto* biocompatibility, we have to take a look at how this impression may have come about. The key building blocks that nature utilizes to assemble into the biopolymers chitin and chitosan are the amino-sugars N-acetyl-glucosamine and N-glucosamine. It would be reasonable (and correct) to also expect that what nature puts together, nature can also disassemble. Therefore, since nature handles the bioavailability of chitin and chitosan so elegantly, by inference the biopolymers are environmentally friendly, biodegradable, and by extension, biocompatible. What is implied (and popularly accepted) is that chitin and chitosan being natural materials are biocompatible and not harmful to humans.

The perplexity has to do with the term biocompatibility. The term biocompatibility as evolved by the community of biomaterials and medical device practitioners in partnership with Regulatory and Standards bodies is explicit with respect to its definition. Specifically, biocompatibility relates to an intended use when applied to biomedical applications and must be demonstrated before the term can be applied to a device and its associated biomaterial(s). This is normally compiled into a portfolio containing all performance and safety studies for a single device. Furthermore, with reference to the biological origin of chitin, what has been implied for a material in nature does not necessarily hold when that material is removed from nature, processed and used for a biomedical application. By removing from nature and processing chitin in a manner not known to nature instantly introduces uncertainty to the notion of environmentally friendly and biodegradable. It follows that biocompatibility of chitin has also to be established in conformance to accepted practices and therefore to label chitin *de facto* biocompatible in all instances may not be in the best interest of advancing chitin for biomedical applications.

How should we progress from here? First, by establishing that the many well-documented studies on chitin and chitosan do support the claims of chitin being biocompatible. Second, to consciously work towards completing the biocompatibility picture for chitin.

Since the early days of chitin research, biomedical applications have been stated as a likely channel that can ultimately be applied in some useful form. Consequently, many studies focused on demonstrating the performance of chitin for the sought after biomedical application as highlighted in Chapter 2. Around the mid-1980's, a more formal emphasis on "core" biocompatibility using cell culture and animal models became evident. Provided below is a snapshot summary of the biocompatibility studies that have been reported for chitin and chitosan from which a perspective of the effort already expended can be obtained. This survey is focused primarily on safety and biological performance studies, the bulk of which are biological evaluations that normally form the core of toxicity evaluations. The reason for focusing on biological evaluations in preference to chemical, mechanical and physical properties is that the latter are more performance-based and as stated above, application dependent. In addition biological evaluations have a direct and immediate impact, as they are good indicators of the response of biological systems especially toxicologically.

4.3 SURVEY OF THE BIOCOMPATIBILITY STUDIES ON CHITIN

Cell culture methods have been developed to elicit both toxicological aspects as well as cell - biomaterial interactions of biomaterials. Schmidt et al investigated the effect by several polysaccharides that included chitin and chitosan on L929 fibroblast cells proliferation including a macrophage activity assay.[2] Chitin gave a low cell yield relative to control at day 3, decreasing further at day 6 indicating chitin did not support cell proliferation. However, cells exposed to chitin were viable, therefore chitin was not cytotoxic. In contrast, chitosan in the form of the lactate salt was found to affect the cell proliferation rate, the effect diminishing somewhat with increasing chitosan concentration to a maximum after which inhibition of cell proliferation resumed. The non cell-proliferative character was attributed to the possibility of chitin-metal ions binding; generating species that in the appropriate concentration stressed cells impeding cell proliferation. Furthermore, chitin was found to have no effect on macrophage activity while chitosan lactate displayed a very inhibitory outcome. Subsequent work by the same group using fungal mycelia studied the proliferation of human L1000 fibroblast cultures.[3] Cell proliferation was found to be dependent on the fungal species that also affected cell morphology, attributed to the chitin content in the fungi. In their later report, the authors related the generation of hydrogen peroxide by the fungal materials as the influencing factor on their cell proliferation results obtained with L929 fibroblast cells.[4]

In a separate cytotoxicity evaluation, chitosan-polyvinyl pyrrolidone hydrogels were placed in contact fibroblast (NIH3T3) and epithelial (SiHa) cells.[5] Both direct contact and MTT (3,4,5-dimethylthazolyl-2)-2,5-diphenyl tetrazolium bromide) assay evaluation on leachables extracts indicated non-cytotoxicity. Other general cell culture evaluations demonstrating primarily the non-cytotoxic character of chitin-based materials include work by Minoura et al where chitosan-polyvinyl-alcohol hydrogels in contact with L929 mouse fibroblast cells showed favorable growth, attachment and proliferation.[6] This effect was more pronounced when the chitosan content in the blend was between 15 to 40% by weight and was superior to collagen.

Guerra et al investigated the influence of chitosan-polyelectrolyte complexes on cell proliferation, lysosomal and mitochondrial activity.[7] The positive cell proliferation behavior for the chitosan-polyacrylic acid complex was attributed to the structure of the complex and its extract being favorable to cells. Cytotoxicity assays using NIH3T3 and HeLa cell lines with MTT and neutral red on chitosan-polyacrylamide hydrogels extracts have also been reported to be non-cytotoxic.[8] Balsinde et al demonstrated that chitosan induced macrophage activity and could have use as a pro-inflammatory stimulant.[9]

Most recently, two studies on the effect of the degree of acetylation on cell behavior have been reported. In one report using ASTM methods to assess cytotoxicity, chitin and chitosan films obtained from shrimp and squids were exposed to the popular L929 mouse fibroblast and BHK21(C13) hamster kidney cells.[10] No growth inhibition zones were found in cell culture samples for all chitosan films and extracts, indicating no cytotoxic effects. However, cells were found to adhere to the more deacetylated samples while the converse was true for the higher acetylated samples, this observation being related to the amount of free amino groups present. In the other study, Domard et al utilized keratinocytes and fibroblasts cells.[11] Again, all the chitosans were found to be cytocompatible to the two cells types regardless of their degree of acetylation. However, the adhesion of keratinocytes was reduced when the degree of acetylation decreased leading to poorer keratinocyte cell proliferation. Fibroblasts displayed high adhesion and poor cell proliferation for all degrees of acetylation. The authors suggest that a suitable low degree of acetylation chitosan would favor wound healing, as this would adhere and control fibroblast proliferation, desirable in wound healing, while permitting keratinocytes, the preferred cell type to proliferate, promoting wound healing.

Performance evaluation using cell culture is equally important. Canine polymorphonuclear cell (PMN) extracts in the presence of chitin or chitosan in vitro have been shown to stimulate the release of leukotriene B4 (LTB4) by Okamoto et al.[12] It was also noted that chitosan induced a marked inflammatory response by an increase in a large number of blood cell materials in peritoneal exudative fluid (PEF). The presence of PMNs, and the release of LTB4, which promotes PMN migration, suggests wound-healing roles by chitin and chitosan. Another example is the assessment of the influence by chitin and its derivatives on the production of cytokine and tumor necrosis factor (TNF), important materials secreted by cells as part of the wound healing process using human umbilical vein endothelial cells (HUVEC) by Mori et al.[13] Chitin and chitosan did not affect the cell proliferation of HUVEC cells but only a N-sulfated derivative stimulated the secretion of cytokines and TNF.

These series of studies focus on the use of *in vitro* cell culture that can be a resource for general cytotoxicity and cytocompatibility studies as well as performance evaluation of chitin/chitosan as wound dressings.

The use of animal model studies is unavoidable in biomedical related fields as they provide a more comprehensive indication of toxicity, and tissue interactions not possible with cell cultures. Nishimura and co-workers who focused on the elicitation of immunological responses in mice and guinea pigs performed some of the first studies.[14] Chitin and chitosan of varying degrees of deacetylation and some water-soluble derivatives prepared as suspensions or solutions were injected into mice and the activation of murine peritoneal macrophages evaluated 3-15 days later. The tumor suppression study was similarly performed, the variation being the inclusion of tumor cells in the chitin solutions. A positive ability to stimulate peritoneal macrophage was interpreted as a favorable indicator of evoking an immune response against tumor and infection. 30% and 70% deacetylated chitins and carboxymethyl-chitin were the best at activating the peritoneal macrophages and 70% deacetylated chitin was best in suppressing tumor growth, while 30% deacetylated chitin exhibited only a moderate effect. These two forms of deacetylated chitins were also found to be strong activators of adjuvant activity, inducing cell-mediated immunity and circulating antibody formation in mice and guinea pigs. Subsequent work related the activation of mouse peritoneal macrophages by carboxymethyl-chitin to the degree of substitution where a high anionic charge was more effective. Finally in this series, 80% deacetylated chitin beads also activated mouse peritoneal macrophages suggesting an immune system stimulating role for chitin in drug delivery systems.

Rao and Sharma have noted the dearth of toxicological data on chitosan and presented, among other studies, their safety studies using animal models.[15] Acute systemic toxicity, rabbit pyrogen and intramuscular implantation tests of chitosan samples were assessed and results showed no toxicity effects. Tanaka et al administered chitin and chitosan orally, parenterally and subcutanoeusly into mice.[16] Mice administered with 1 or 5 mg of chitin or chitosan intraperitoneally once every 2 weeks over a 12-week period, were assessed as normal based on general observations. Only mice given the higher dosage chitosan showed accompanying weight loss. However, histological results revealed that in chitin treated mice, macrophages with hyperplasia in the mesenterium abound, and foreign-body giant cell type polykaryocytes were present in the spleen. Similar histological findings were obtained for chitosan. Furthermore, when administered subcutaneously, chitosan was found not to invoke a significant cellular response suggesting the biodegradation of chitosan *in vivo* might occur very slowly. The results implicate that chitin and chitosan may be unacceptable for some medical applications, and warranted further investigations, especially for chitosan.

The fate of chitin in tissue is important as it may give rise to complications such as thrombosis.[17] To address this issue, Minami et al investigated the systemic effect of chitin by administering chitin polymer and oligomer as physiological saline suspensions/solutions into a dog model and evaluated the white blood cell (WBC) count. No systemic effect was noted with the oligomer. Chitin suspensions induced a significant decrease in WBC count attributed to the particulate interaction of chitin with blood. In a follow-up study, where administration was subcutaneous, chitin and chitosan suspensions, and chitosan oligomer solution were used with latex for comparison.[18] Minimal responses were obtained for chitin, chitosan oligomers and latex control, but chitosan elevated the WBC count, the extent being dose dependent. The results indicate that chitosan induced host responses that may be useful in against surgical and anesthetic stress. Further studies by the same group have shown chitin and chitosan stimulates the complement activity by the alternative pathway.[19] This activity was dependent on the degree of acetylation as well as the method of chitin and chitosan preparation, homogeneously or heterogeneously.[20]

In the use of a feline *in vivo* model in wound healing, Kojima et al investigated the formation of granulation tissue on wounds stimulated by chitin and chitosan containing implants after a 14-day implant period.[21] Granulation tissue excised from the wound mediated by the chitin containing implant was thin and displayed minimal foreign body reaction. In comparison, the chitosan exposed wound site was found to produce excessive granulation tissue, extending into the surrounding tissue. The results suggest better stimulation of the wound site by chitin to produce preliminary collagen followed by digestion of this collagen as the wound site reorganized the collagen assembly while chitosan elicited a greater and continuous inflammatory response unfavorable for proper wound healing. The authors suggested a role with the poor biodegradability of chitosan that warranted further work to understand these results.

Tabrizian et al investigated the response of L929 fibroblast cells by MTT assay of chitosan-xanthan polyionic complex hydrogels.[22] The particles and their extracts did not elicit any cytotoxic effects. Particles of the hydrogel stimulated J-774 macrophage cells to produce TNF *in vitro* indicating macrophage activation. *In vivo* assessment using a rat model showed fragments of the hydrogel after 12 weeks suggesting eventual biodegradation.

The data compiled above support the claim that chitin is biocompatible both in terms of a low level of toxicity to humans and performance in various applications. This conclusion warrants the pursuit of making chitin an acceptable biomaterial.

4.4 OUTSTANDING MATTERS

The preceding discussion on the biocompatibility of chitin and chitosan is a smidgen of the innumerable accounts on this topic found in the literature. While extensive, most accounts are non-specific or incomplete. For a biomaterial to be demonstrated to be biocompatible, corroborative data based on other considerations are necessary to complete a comprehensive profile. As alluded-to before, this can only be meaningful when:

1.	The specific biomedical application i.e. device is identified.

2.	The chitin material to be evaluated has a production profile that is verifiable i.e. produced in a controlled procedure and important characterization information known.

3.	A proper assessment of the type of biocompatibility studies required. That is to say a reasonable assortment of experiments necessary to demonstrate safety and that the device and biomaterials will perform its intended purpose. No overkill is necessary.

The necessary line up of biocompatibility experiments is well documented and it is the responsibility of the device manufacturer to sort out the appropriate list needed to achieve their goals i.e. bring the device to market. With regard to the chitin material, the question is whether the production and characterization of these materials are at a stage that permits easy verification for biomedical use.

4.5 REFERENCES

[1] Concise Encyclopedia of Medical and Dental Materials. D.F. Williams, Ed, Pergamon Press, Oxford, UK. 1990. 52

[2] R.J. Schmidt, L.Y. Chung, A.M. Andrews, O. Spyratou, T.D. Turner, Biocompatibility of wound management products: A study of the effects of various polysaccharides on Murine L929 fibroblast proliferation and macrophage respiratory burst. J. Pharmaceutics and Pharmacology 45 (1993) 508-513

[3] L.Y. Chung, R.J. Schmidt, P.F. Hamlyn, B.F. Sagar, A.M. Andrews, T.D. Turner, Biocompatibility of potential wound management products: Fungal mycelia as a source of chitin/chitosan and their effect on the proliferation of human F1000 fibroblasts in culture. J. Biomedical Material Research 28 (1994) 463-469

[4] L.Y. Chung, R.J. Schmidt, P.F. Hamlyn, B.F. Sagar, A.M. Andrews, T.D. Turner, Biocompatibility of potential wound management products: Hydrogen peroxide generation by fungal chitin/chitosans and their effects on the proliferation of murine L929 fibroblasts in culture. J. Biomedical Materials Research 39 (1998) 300-307

[5] M. Risbud, A. Hardikar, R. Bhonde, Growth modulation of fibroblasts by chitosan-polyvinyl pyrrolidone hydrogel: Implications for wound management? J. Bioscience 25 (2000) 25-31

[6] N. Minoura, T. Koyano, N. Koshizaki, H. Umehara, M. Nagura, K-I. Kobayashi, Preparation, properties, and cell attachment/growth behavior of PVA/chitosan blended hydrogels. Materials Science and Engineering C6 (1998) 275-280

[7] G.D. Guerra, P. Cerrai, M. Tricoli, S. Maltini, R. Sbarbati del Guerra, *In vitro* cytotoxicity testing of chitosan-containing polyelectrolyte complexes. J. Materials Science: Materials in Medicine 9 (1998) 73-76

[8] M.V. Risbud, R.R. Bhonde, Polyacrylamide-chitosan hydrogels: *In vitro* biocompatibility and sustained antibiotic release studies. Drug Delivery 7 (2000) 69-75

[9] I.D. Bianco, J. Balsinde, D.M. Beltramo, L.F. Castagna, C.A. Landa, E.A. Dennis, Chitosan-induced phospholipase A_2 activation and arachidonic acid mobilization in $P388D_1$ macrophages. FEBS letters 46 (2000) 292-294

[10] M. Prasitsilp, R. Jenwithisuk, K. Kongsuwan, N. Damrongchai, P. Watts, Cellular responses to chitosan *in vitro*: The importance of deacetylation. J. Materials Science: Materials in Medicine 11 (2000) 773-778

[11] C. Chatelet, O. Damour, A. Domard, Influence of the degree of acetylation on some biological properties of chitosan films. Biomaterials 22 (2001) 261-268

[12] Y. Usami, Y. Okamoto, T. Takayama, Y. Shigemasa, S. Minami, Chitin and chitosan stimulate canine polymorphonuclear cells to release leukotriene B_4 and prostaglandin E_2. J. Biomedical Materials Research 42 (1998) 517-522

[13] T. Mori, Y. Irie, S-I. Nishimura, S. Tokura, M. Matsuura, M. Okumura, T. Kadosawa, T. Fujinaga, Endothelial cell responses to chitin and its derivatives. J. Biomedical Materials Research: Applied Biomaterials 43 (1998) 469-472

[14] K. Nishimura, S. Nishimura, N. Nishi, I. Saiki, S. Tokura, I. Azuma, Immunological activity of chitin and its derivatives. Vaccine 2 (1984) 93-99; K. Nishimura, S. Nishimura, N. Nishi, F. Numata, Y. Tone, S. Tokura, I. Azuma, Adjuvant activity of chitin derivatives in mice and guinea-pigs. Vaccine 3 (1985) 379-384; S-i. Nishimura, N. Nishi, S. Tokura, Bioactive chitin derivatives: Activation of mouse peritoneal macrophages by O-(carboxylmethyl)chitins. Carbohydrate Research 146 (1986) 251-258; K. Nishimura, S-i. Nishimura, H. Seo, N. Nishi, S. Tokura, I. Azuma, Effect of

multiporous microspheres derived from chitin and partially deacetylated chitin on the activation of mouse peritoneal macrophages. Vaccine 5 (1987) 136-140

[15] S.B. Rao, C.P. Sharma, Use of chitosan as a biomaterial: Studies on its safety and hemostatic potential. J. Biomedical Materials Research 34 (1997) 21-28

[16] Y. Tanaka, S Tanioka, M. Tanaka, T. Tanigawa, Y. Kitamura, S. Minami, Y. Okamoto, M. Miyashita, M. Nanno, Effects of chitin and chitosan particles on BALB/c mice by oral and parenteral administration. Biomaterials 18 (1997) 591-595

[17] S. Minami, R. Mura-e, Y. Okamoto, T. Sanekata, A. Matsuhashi, S. Tanioka, Y. Shigemasa, Systemic effect of chitin after intravenous administration to dogs. Carbohydrate Polymers 33 (1997) 243-249

[18] S. Minami, M. Masuda, H. Suzuki, Y. Okamoto, A. Matsuhashi, K. Kato, Y. Shigemasa, Subcutaneous injected chitosan induces systemic activation in dogs. Carbohydrate Polymers 33 (1997) 285-294

[19] S. Minami, H. Suzuki, Y. Okamoto, T. Fujinaga, Y. Shigemasa, Chitin and chitosan activate complement via the alternative pathway. Carbohydrate Polymers 36 (1998) 151-155

[20] Y. Suzuki, Y. Okamoto, M. Morimoto, H. Sashiwa, H. Saimoto, S-i. Tanioka, Y. Shigemasa, S. Minami, Influence of physico-chemical properties of chitin and chitosan on complement activation. Carbohydrate Polymers 42 (2000) 307-310

[21] K. Kojima, Y. Okamoto, K. Miyatake, Y. Kitamura, S. Minami, Collagen typing of granulation tissue induced by chitin and chitosan. Carbohydrate Polymers 37 (1998) 109-113

[22] F. Chellat, M. Tabrizian, S. Dumitriu, E. Chornet, P. Magny, C-H. Rivald, L. Yahia, *In vitro* and *in vivo* biocompatibility of chitosan-xanthan polyionic complex. J. Biomedical Materials Research 51 (2000) 107-116

CHAPTER 5: THE SOURCES AND PRODUCTION OF CHITIN

5.1 SOURCES OF CHITIN

In the production of a biomaterial, certain basic requirements must be met, such as high purity and consistency from batch to batch, before it can be called a biomedical grade material. This is an especially important concern in the production of chitin as its biological origin equates to a wider variation in the source raw material such as the calcium and protein content in the shells differing with sex, age and habitat of the animal.[1] Furthermore, it is essential to recognize that chitin can be obtained from several marine and plant sources. The choice of the source affects the extraction and isolation methods of chitin from its biological source that in turn affects the number of steps and types of chemicals used. All these considerations influence the quality of the final processed chitin. These concerns are addressed below.

Chitin is widely distributed both in the animal and plant kingdom.[2] In animals, the most readily associated sources are in the shells of crustaceans and mollusks, the backbone of squids and the cuticle of insects. Chitin's role in the shell and cuticle is as a structural component that contributes strength and protection to the organisms. In crustaceans or more specifically shellfish, chitin is found as a constituent of a complex network with proteins onto which calcium carbonate deposits to form the rigid shell. The interaction between chitin and protein is very intimate, with covalent bonding present, and in essence, is a polysaccharide-protein complex.[3] **Chitosan** is not native to animal sources and is normally obtained by the deacetylation of shellfish-derived chitin using sodium hydroxide.

In the plant kingdom, **chitin** is present in the algae, commonly known as marine diatoms, protozoa and the cell wall of several fungal species.[4] Chitin from the diatom spines such as *Cyclotella cryptica* and *Thalassiosira fluviatilis* are the only form reported to be 100% poly-N-acetyl-glucosamine that is not associated with proteins and is termed **chitan**.[5] A small number of fungal strains are known to produce **chitosan** in preference to chitin.[6]

In fungi, chitin is the principal fibrillar polymer in the innermost layer of the cell wall with septate mycelium, a common feature in most of the higher fungi.[7] Fungal chitin occurs as randomly oriented microfibrils intertwined and embedded in an amorphous matrix to provide the framework in cell wall morphology. Furthermore, fungal chitin is covalently linked to other molecules such as glucans, a requirement of their biological role. Fungal cell wall is approximately 80% polysaccharide, with proteins and lipids constituting the remainder.[8]

The chitin content of fungi varies from a low of ~0.45% in yeasts, while filamentous fungi species with high proteolytic activity can contain up to 10-40% chitin.[9] Values as high as 58% are found in aquatic molds such as *Allomyces macrogynus* and 91% in the species *Phycomyces blakesleeanus* have also been reported.[10] Despite their high chitin content, *Allomyces* are not favored as candidates for chitin production from fungi because there have been indications that aquatic molds are not readily propagated on a large-scale.

5.2 CHITIN PRODUCTION

While chitin is available from a myriad of animal and plant sources, traditionally, only the animal source has been exploited on a commercial scale. Table 5.1 summarizes the total global estimates for chitin from commercially exploitable marine sources.[11] It should be noted that the amount of chitin available from shellfish is subject to seasonal supply and their demand as a luxury food item.

Resource	Landings (MT)	Potential waste (MT)	Estimated waste (MT)	Dry waste (MT)	Chitin content (MT)
Shrimp	2 647 345	1 058 938	710 000	177 500	44 375
	(40)*		(0.25)**	(0.25)***	
Squid	1 991 094	389 219	99 531	24 882	1 244
	(20)*		(0.25)**	(0.05)***	
Crabs	1 348 323	943 826	482 744	144 823	28 964
	(70)*		(0.30)**	(0.20)***	
Oyster/clam	2 547 287	1 783 100	304 948	274 453	12 350
	(70)*		(0.90)**	(0.045)***	
Krill	232 700	93 080	93 080	23 270	1 629
	(40)*		(0.25)**	(0.07)***	
Total	8 766 749				88 652

*   Multiplication factor for calculating waste
**   Multiplication factor for calculating dry waste
*** Multiplication factor for calculating chitin[12]

Table 5.1:   Global estimates of potential chitin from shellfish sources[11]

The dry shells of crabs, lobsters and shrimps contain 20-40% chitin, 30-40% of recoverable proteins and 20-30% of calcium carbonate. They are the most readily accessible sources as waste from the seafood processing industry and essentially, all the presently commercially produced chitin are derived from them. Realistically, the available chitin supply is about 75000 metric tons (MT) from shellfish from a seafood processing waste of approximately 2.0 x10^6 metric tons annually. Chitin from shellfish has been shown to be economically viable together with protein, pigment and mineral recovery as by-products. The economic contribution from protein and calcium carbonate to the commercial viability of chitin production cannot be understated.

Krill has a lower chitin content of <10% and the primary product obtained is protein.[13] Clam and oysters contain a large inorganic component of up to 90% dry weight. The estimated total chitin tonnage of 15000 for krill, squid and clams/oysters make these sources commercially less viable for the present. Even if the overall estimate of Brine at 150 x 10^3 tons (twice the amount stated above) is utilized, it is likely that the production of chitin from shellfish sources will have a finite limit whose volume may not meet demand in the not too distant future.[14] Interestingly, Brine has further indicated that from the supply viewpoint, implying cost considerations, chitin can only be a specialty polymer. In other words, while chitin is known to be the second most abundant polysaccharide in nature, the reality is that this abundance is not directly transformed into a plentiful chitin supply. This will restrict the use of chitin to situations where the cost merits its use, as in biomedical applications.

One way of circumventing this potential supply question is for fungal chitin to take up the slack in chitin supply. It has been estimated that fungi could provide 3.2×10^4 metric tons of chitin annually and can be potentially limitless if required.[15] Chitin from fungal mycelia is seen as an especially important alternative to shellfish sources. There are several benefits of using fungal mycelia as an alternative to shellfish. Fungal biotechnological processes are rapid and can be organized in a closed or semi-closed technological circuit to comply with modern ecological requirements. This would provide chitin all year round obviating the uncertainty of shellfish supply that is subject to seasonal supply fluctuations. Inexpensive raw material such as lignocellulose wastes can also be used as carbon sources to increase cost effectiveness. Furthermore, advanced fungal biotechnologies leave little wastes compared to chemical processing of shellfish and can produce several useful end products e.g. organic acids. The raw chitin material from fungi is relatively consistent in composition. Finally, recent increases in the levels of heavy metals and radioactivity in shellfish from pollution coupled with their perceived allergenic problems render chitin from shellfish sources unfavorable for biomedical applications. The trend towards biomedical applications provides a further incentive for the production of fungal chitin.

5.3 ISOLATION OF CHITIN

The primary biological function of crustacean chitin and fungal chitin is to provide a structural scaffold in support of the animal exoskeleton or fungal cell wall. However, this function is fulfilled differently and is reflected in their physicochemical properties. Shellfish chitin is an exoskeletal component in a complex network containing proteins and minerals $(CaCO_3)$ while the main components in the complex network of fungal chitin are other polysaccharides such as α- and β-glucan, mannan and cellulose. Shellfish chitin is more crystalline and chemically more stable while fungal chitin is "soft" and less crystalline. Shellfish chitin is more acetylated compared to fungal chitin that has a lower degree of acetylation. Therefore, these differences have impact on the extraction processes utilized to produce chitin.

It is evident that regardless of whether the source of chitin is animal or plant, there is an intimate link between the biopolymer with the biological system it is found. In particular, chemical bonding with proteins and the composite-like link with calcium carbonate make the accessibility of the biopolymer from shellfish intricate. Similarly, the chemical bonding with glucan and proteins in the fungal cell wall have to be severed in order for chitin to be isolated. As the use of chitin in its native form is not known, the close association of chitin with its biological surroundings in both animal and plant sources implies that some form of separation is necessary. This posed a considerable challenge in the isolation of the biopolymer. The methods that have been devised have ultimately been based on understanding the properties of chitin and its native environment.

5.3.1 Isolation of Chitin from Shellfish

The process of isolating chitin from shellfish involves the step-by-step removal of the two major constituents of the shell, the intimately associated proteins by deproteinization and inorganic calcium carbonate by demineralization, together with small amounts of pigments, lipids and trace-metals leaving chitin as the final residue.[16] Many methods have been proposed and used over the years.

The isolation of chitin begins with the selection of shells. For crabs and lobsters, this selection has important bearing on the subsequent quality of the final isolated material. Ideally, shells of the same size and species are chosen. For shrimp shells, the comparatively thin walls make recovery of chitin easier. Cleaning and drying of the shells followed by thorough crushing is the next step in the process. The small shell pieces are treated with dilute hydrochloric acid to remove calcium carbonate. Proteins as well as other organic impurities are removed by an alkali treatment (20% sodium hydroxide). Pigments, primarily carotenoids are removed by extraction with ethanol or acetone after the demineralization process. A typical procedure is outlined in Figure 5.1 (left-half of Figure) to give raw chitin.

5.3.1.1 Demineralization

Demineralization as the term suggests is the removal of minerals, primarily calcium carbonate. A mineral free chitin i.e. very low ash content chitin would be required for applications that have very low impurity tolerance. Demineralization is readily achieved because it involves the decomposition of calcium carbonate into the water-soluble calcium salts with the release of carbon dioxide. The most used reagent is dilute hydrochloric acid (HCl) that produces water-soluble calcium chloride ($CaCl_2$). Some of the shortcomings of using HCl are that it is relatively expensive, large volumes of $CaCl_2$ solution must be disposed of and being a strong acid, can cause some hydrolysis of the chitin chains thereby

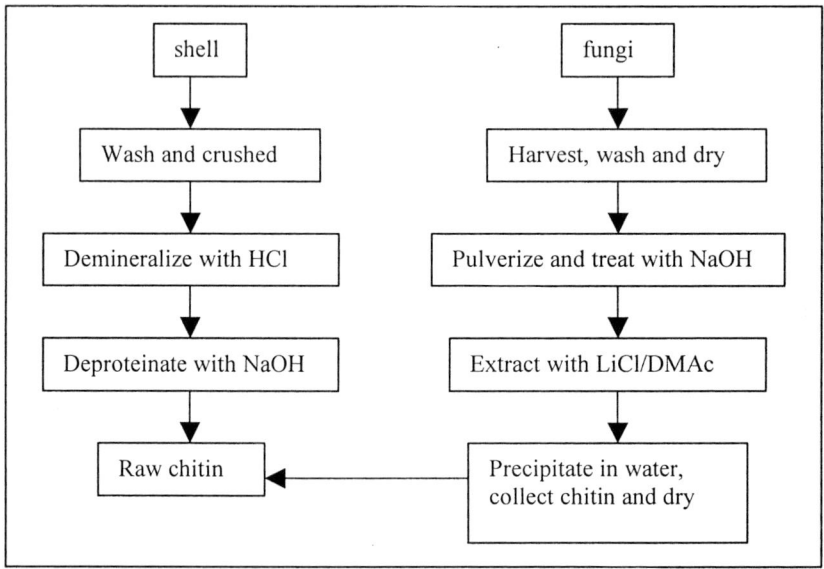

Figure 5.1:   Separation and isolation of chitin from shells and fungi

reducing the molecular weight of the biopolymer. Performing demineralization at room temperature or lower can minimize this depolymerization of chitin chains. Other milder forms of demineralization include EDTA extraction and microbial demineralization where a low pH is obtained in these fermentations. Properly performed, the calcium content in as received chitin from traditional commercial sources is less that 1%.

5.3.1.2 Deproteinization

While demineralization is straightforward, deproteinization is not. In deproteinization, covalent chemical bonds have to be destroyed between the chitin-protein complex. This is achieved with some difficulty especially if performed heterogeneously utilizing chemicals that will also depolymerize the biopolymer. The complete removal of protein, possible from shellfish sources, is especially important for biomedical applications, as a percentage of the human population is allergic to shellfish, the primary culprit being the protein component.

Chemical methods were the first approach used in deproteinization. A wide range of chemicals have been tried as deproteinization reagents including $NaOH$, Na_2CO_3, $NaHCO_3$, KOH, K_2CO_3, $Ca(OH)_2$, Na_2SO_3, $NaHSO_3$, $CaHSO_3$, Na_3PO_4 and Na_2S. $NaOH$ is the preferred reagent and typically a $1M$ $NaOH$ solution is used with variations in the temperature and duration of treatment parameters. The use of $NaOH$ invariably results in partial deacetylation of chitin and hydrolysis of the biopolymer that lowers the molecular weight of chitin.[17] Properly carried out, the protein content in as received chitin from traditional commercial sources is around 1%.

In seeking to raise the deproteinization efficiency and an alternative to the harsh chemical treatment other methods have emerged and are being investigated. The use of proteolytic enzymes such as pepsin, papain or trypsin has been shown to minimize deacetylation and depolymerization in the chitin isolate. Other proteolytic enzymes such as tuna trypsin, Rhozyme-62, cod trypsin and bacterial proteinase have also been demonstrated to remove proteins from crustacean shells.[18] It must be noted that the efficiency of enzymatic methods is inferior to chemical methods with approximately 5% residual protein typically still associated with the chitin isolate. The final chitin isolate will have to be treated with an additional $NaOH$ treatment, albeit under milder conditions and for a shorter duration to achieve the purity obtained with $NaOH$ deproteinization only. Finally, it should be noted that commercially purified enzymes could be costly.

The cost of using enzymes can be lowered if deproteinization is performed in the presence of microorganisms that secrete proteolytic enzymes during fermentation. An added benefit of microbial deproteinization for chitin isolation is a reduced environmental impact. For example, the bacterial fermentation of lactic acid in the presence of crustacean shells lowers the pH of the medium to approximately pH 4. The lowered pH enhances the hydrolysis of proteins that is harvested and used for animal feed leaving behind the chitin residue. Bustos and Healy have found that chitin obtained by deproteinization of shrimp shell waste using proteolytic microorganisms such as *Pseudomonas maltophilia*, *Bacillus subtilis*, *Streptococcus faecium*, *Pediococcus pentosaseus* and *Aspergillus oryzae*, had a higher molecular weight compared to chemically prepared chitin.[19] Teng has also demonstrated that the concurrent production of chitin from shrimp shells and fungi in a one-pot fermentation process where proteases from the fungi hydrolyzes the protein into amino acids that in turn act as a nitrogen source for fungal growth is possible.[20]

Microbial fermentation does have its drawbacks such as requiring a longer processing time compared to chemical methods and a poorer accessibility of certain portions of the shellfish to proteases. Provided some of the limitations of microbial fermentation are addressed and resolved, the practicality of microbial deproteinization on an industrial scale holds promise.

The removal of these two major components, calcium carbonate and proteins gives raw chitin of reasonable quality. However, it must be noted that the quality of the final chitin obtained is first directly dependent on the starting raw material that needs to be consistent. For example, it would be preferable for all the shells obtained to be from similar aged crustaceans. The manner in which the shells are gathered, cleaned, dried and powdered will also affect the final quality as this determines the amount of impurities and accessibility of the shells to the chemicals or enzymes. Raw chitin can be further processed depending on end-use requirements as in biomedical applications. It must be noted that in the recovery of any bioresource, the elimination of the last fraction of unwanted biological components is difficult and may not be necessary. There is insignificant economic benefit in doing so especially if it contributes little to increasing purity and degrades the biopolymer but may be warranted for biomedical applications.

5.3.1.3 Deacetylation of Chitin into Chitosan

The deacetylation of chitin into chitosan is usually achieved by treating chitin with 50% NaOH at 95°C for 3 hours, cooling down decanting off the NaOH and washing with water until neutral pH. This procedure is normally repeated twice. Finally the chitosan is extracted with 2% acetic acid solution, filtered and precipitated in distilled water to give purified chitosan that is dried and stored.

A reference method for the isolation of chitin and chitosan by No and Meyers is found in the Chitin Handbook.[21]

5.3.2 Physical Appearance

Well processed, with a reasonable degree of purity chitin and chitosan are white and odorless powders. They can be used for most non-biomedical applications as outlined in Table 1.1 of Chapter 1. However, very often as received materials from vendors in the form of flakes or powder range from deep to light yellow in color and can possess a decaying seafood odor. Very crude raw materials even contain fragments of shells that have been poorly processed! Figure 5.2 shows the typical flake and powder forms of chitin together with freeze-dried fungal mycelia from which fungal chitin are isolated.

As received materials are normally further processed prior to use. In the author's laboratory, as received chitin material is first soaked in aqueous 5% NaOH solution at room temperature for 7 days at a stirring speed of 200rpm using a mechanical stirrer. This is followed by washing of the treated chitin with deionized water to neutral, followed by treatment with 1M HCl for 1 hour and finally washed to neutral and dried. This procedure has been shown to provide chitin of reasonable quality adequate for our research purposes. A Standard protocol prepared in the Author's laboratory for reference is attached as Appendix 1. The chitin obtained from this process readily dissolves in the standard 5%LiCl/DMAc (lithium chloride/N, N-dimethylacetamide) solvent system to give a clear chitin solution. The chitin can also be deacetylated into chitosan using standard procedures as referenced above.

5.3.3 Isolation of Chitin from Algae and Fungi

The isolation of chitin from fungal cell wall involves deproteinization and lipid extraction. The mineral content in fungi is minimal and is normally not an issue. This is a definite advantage in providing a biomedical grade material in contrast to shellfish sources. Deproteinization is achieved with NaOH or enzymes as detailed above. Lipid is usually removed using organic solvents such as acetone. The residue however is not pure chitin but the fungal-glucan complex. The glucan component can be hydrolyzed enzymatically at an added cost. Alternatively, the fungal-glucan-chitin complex can be used as it is, so far without allergenic complications.[22] However, the implications of using this complex has not been further elaborated and at present is premature to conclude whether the fungal-glucan-chitin complex can be used as a biomedical material, where purity is concerned.

A typical procedure used in the Author's laboratory is shown in Figure 5.1 (right-half of Figure). The solvent 5% LiCl/DMAc is used to extract the chitin-glucan complex and subsequently precipitated in water to give the chitin-rich complex.

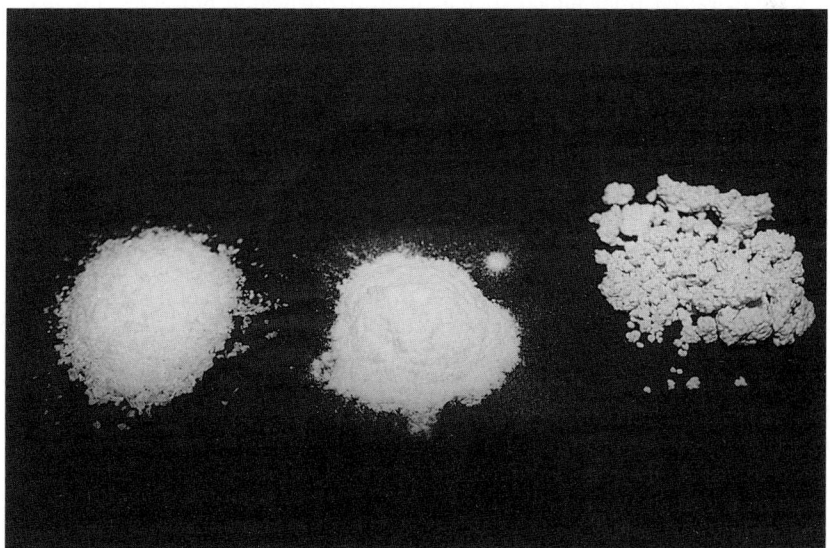

Figure 5.2:   Chitin flakes, powder and freeze-dried fungal mycelia

The isolation of chitosan from fungal mycelia would involve similar procedures of deproteinization. A 2% acetic acid solution is used to extract the chitosan that is subsequently filtered and reprecipitated in water to give fungal chitosan.[23] The processes that have been applied to-date to isolate chitin from fungal sources have in general been successful at recovering chitin and chitosan of reasonable quality.[15]

Perhaps the purest form of chitin or more correctly chitan is obtained from the algal diatoms. Vournakis et al have presented in a U.S. patent a process for growing and recovery of chitan for biomedical use.[24] Mechanical force method and chemical/biological methods using similar processes as described above are used to produce pure chitan. The status of chitan using this process awaits formal introduction but as has been noted earlier in Chapter 2, a biomedical product based on material derived from this process has already appeared.

5.4 CONSEQUENCE

There are several consequences in reviewing the types of resources pertaining to the supply of chitin as they have important implications in turning chitin into a biomedical material. First, the common claim that chitin is a material obtained from a renewable resource deserves some scrutiny. It is absolutely correct to state that as far as nature is concerned, chitin is a renewable resource. However, in the exploitation of chitin, this statement has to be treated with caution. The amount of utilizable chitin is very much less than the billions of ton estimated to be available in the biosphere. This has been attributed to the non-economic viability of many such ventures such as too little chitin content in certain marine animals, the pooling of inadequate shellfish waste in a particular region, etc. However, if the chitin and chitosan were extracted from fungal sources using fermentation processes, the reality of a material from a renewable resource may be attained.

However, the potential limitless supply of chitin from fungi must also be approached with caution, as this has not been demonstrated on a mass production scale. Whether there will be problems on scaling up and whether it will be economically viable, as chitin content is lower in fungi compared to shellfish, is unresolved. Furthermore, as with shellfish waste, there is no avoiding the realization that fungal fermentation must be performed in tandem with a process that provides the additional economic incentive such as producing organic acids.[4] In addition, the stringent requirements of biomedical grade chitin can impose further costs on production that may be prohibitive and has to be resolved if chitin is to make it as a biomaterial. While fungal chitin is not spared complications, it should be easier to establish a quality material. Not much is known about the allergenic properties of chitin derived from fungal sources but the recovery of chitin should ensure that minimal protein is associated. The author advocates that the production of fungal chitin should be considered seriously but ultimately, costs considerations will dictate.

Second, in the present situation, chitin (and chitosan) from shellfish sources will have residual protein and possibly trace minerals while fungal-chitin will have associated glucan. The decision for further processing to reach a biomedical grade has implications on the final cost of the chitin isolate. So far, there is one known source of ultrapure chitosan produced under stringent Good Manufacturing Practices (GMP).[25] It is interesting to note that in 1978, the price for reasonably processed chitosan was estimated at US$2 per pound, roughly equivalent to US$4.40 per Kilogram. The 2001 price for a kilogram of ultrapure grade chitosan is a staggering US$40000 per kg![26]

Finally, it is essential to note that the various chemical and/or microbial treatments under varying conditions of time, temperature and a mix of other environments will affect the biopolymer in terms of its chemical properties as well as purity.

Recently consensus seeking has been put forward to address this issue.[27] Standardization of the processing parameters giving rise to various grades and therefore purity of chitin must be a key priority. Only then can each and every step in the recovery of chitin be evaluated stringently in order to produce a grade of chitin suitable for biomedical purposes.

The shift down the road towards emphasizing the use of chitin as a biomedical material as the higher value-add application will raise the question of whether other established applications would be compromised.

5.5 REFERENCES

[1] P.M. Perceval, The ecnomics of chitin recovery and production. in Proceedings of the First International Conference on Chitin/Chitosan, R.A.A. Muzzarelli, E.R. Pariser, eds., MIT Sea Grant Program, Cambridge, MA, USA, 1978. 45-53

[2] G.A.F. Roberts, Chitin Chemistry. Macmillan Press Ltd, UK, 1992. Chapter 1.

[3] M.N. Horst, A.N. Walker, E. Klar, The pathway of crustacean chitin synthesis. in The crustacean integument: Morphology and biochemistry, M.N. Horst, J.A. Freeman, eds., CRC, Boca Raton, FL, USA, 1993. 113-149

[4] E.P.Feofilova, D.V. Nemtsev, V.M. Tereshina, V.P. Kozlov, Polyaminosaccharides of mycelial fungi: New biotechnological use and practical implications (review). Applied Biochemistry and Microbiology, 32(5) (1996) 437-445

[5] J. McLachlan, A.G. McInnes, M. Falk, Studies on the chitan (Chitin: Poly-N-acetylglucosamine) fibers of the diatom *thalassiosira fluviatilis hustedt*. 1. Production and isolation of chitan fibers. Canadian Journal of Botany, 43 (1965) 707-713

[6] S. Arcidiacono, S. J. Lombardi, D.L. Kaplan, Fermentation, processing and enzyme characterization for chitosan biosynthesis by *Mucor Rouxii*. in Chitin and chitosan: Sources, chemistry, biochemistry, physical properties and applications, G. Skjåk-Bræk, T. Anthonsen, P. Sanford, eds., Elsevier Applied Science, England, UK., 1989. 319-332

[7] M.J. Carlile, S.C. Watkinson, The fungi. Academic Press, London, UK, 1994. Chapter 2.

[8] S. Bartnicki-Garcia, Cell wall chemistry, Morphogenesis, and taxonomy of fungi. Annual Review of Microbiology 22 (1968) 87-108

[9] J. Ruiz-Herrera, The distribution and quantitative importance of chitin in fungi. in Proceedings of the First International Conference on Chitin/Chitosan, R.A.A. Muzzarelli, E.R. Pariser, eds., MIT Sea Grant Program, Cambridge, MA, USA, 1978. 11-21

[10] J. Ruiz-Herrera, Chitin and chitosan. in Fungal cell wall: structure, synthesis, and assembly, CRC Press, Boca Raton, FL, USA, 1992. 89-117

[11] S. Subasinghe, The development of crustacean and mollusc industries for chitin and chitosan resources. in Chitin and Chitosan, The versatile environmentally friendly modern materials , M.B. Zakaria, W.M.W. Muda, M.P. Abdullah, eds., Ampang Press , K.L., Malaysia, 1995. 27-31

[12] G.G. Allan, J.R. Fox, N. Kong, A critical evaluation of the potential sources of chitin and chitosan. in Proceedings of the First International Conference on Chitin/Chitosan, R.A.A. Muzzarelli, E.R. Pariser, eds., MIT Sea Grant Program, Cambridge, MA, USA, 1978. 64-78

[13] S. Nichol, Life after death for empty shells. New Scientist, 129 (1991) 36-38

[14] C.J. Brine, Chitin: Accomplishments and perspectives. in Chitin, chitosan and related enzymes, J.P. Zikakis, ed., Academic Press, Orlando, FL, USA 1984. xvii-xxiv

[15] D. Knorr, Recovery and utilization of chitin and chitosan in food processing waste management. Food Technology 45 (1991) 114-122

[16] K. Shimahara, Y. Takiguchi, Preparation of crustacean chitin. Methods in Enzymology 161 (1988) 417-423

[17] C.J. Brine, P.R. Austin, Chitin variability with strains and method of preparation. Comparative Biochemistry and Physiology 69B (1981) 283-286

[18] G.M. Hall, S.D. Silva, Lactic acid fermentation of shrimp (*Penaus Monodon*) waste for chitin recovery. Biotechnology (1994) 633-638

[19] R.O. Bustos, M.G. Healy, Microbial deproteinization of waste prawn shells. in 2nd International Symposium on Environmental Biotechnology, 1994. 13-15

[20] W.L. Teng, E. Khor, T.K. Tan, L.Y. Lim, S.C. Tan, Concurrent production of chitin from shrimp shells and fungi. Carbohydrate Research 332 (2001) 305-316

[21] H.K. No, S.P. Meyers, Preparation of chitin and chitosan. Chitin Handbook, R.A.A. Muzzarelli, M.G. Peters, eds., Atec Edizioni, Italy 1997. p475-489

[22] C-H Su, C-S Sun, S-W Juan, C-H Hu, W-T Ke, M-T Sheu, Fungal mycelia as the source of chitin and polysaccharides and their applications as skin substitutes. Biomaterials 18 (1997) 1169-1174

[23] S.C. Tan, T.K. Tan, S.M. Wong, E. Khor, The chitosan yield of zygomycetes at their optimum harvesting time, Carbohydrate Polymers, 30, (1996) 239-242

[24] J.N. Vournakis, S. Finkielzstein, E. R. Pariser, M. Helton, Poly-β-1$\rightarrow$4-N-Acetylglucosamine. U.S. Patent 5,623,064. April 22, 1997.

[25] M. Dornish, A. Hagen, E. Hansson, C. Pecheur, F. Verdier, Ø Skaugrud, Safety of Protasan™: Ultrapure chitosan salts for biomedical and pharmaceutical use. Advances in Chitin: II, A. Domard, G.A.F Roberts, K.M. Vårum, eds., Jacques Andre Publishers, Lyon, France, 1997. 664-670

[26] Price as quoted for PROTASAN™ UP grade at US$40 per gram with a minimum order of 10 gram (not adjusted for bulk order) from Pronova Biomedical a.s.

[27] W. Stevens, Production of chitin and chitosan: Refinement and sustainability of chemical and biological processing. 8th International chitin and chitosan conference, 2000. Abstract: K-5, 79

CHAPTER 6: THE STRUCTURAL PROPERTIES OF CHITIN AS IT IS KNOWN TODAY

6.1 INTRODUCTION

The differences between shellfish and fungal chitin reveal that the choice of chitin source and its production methods are an important determining factor of the structural properties of chitin. The structural properties in turn dictate the processing methods, characterization techniques that can be used, types of applications possible and the performance of the final form. A thorough knowledge and understanding of these structural properties is necessary to manipulate chitin to maximize exploitation of its potential.

6.2 SOLID-STATE STRUCTURE

Chitin is now known to have three polymorphic solid-state forms designated as α-, β- and γ-chitin. α-Chitin is the most abundant form, found in shellfish and also in fungal cell wall, although the comparatively lower degree of acetylation of chitin in fungi makes it less rigid. In the solid state, adjacent α-chitin chains are organized in an anti-parallel configuration in the c direction [Figure 6.1].[1] This sets the stage for α-chitin to have a highly ordered crystalline structure with extensive hydrogen bonding that gives rise to the rigid and intractable physical properties of the biopolymer. The pendant N-acetyl functionality plays a major role as it presides over the extensive inter and intra chain N-H·····O=C hydrogen bonds through the C-2 acetamido linkages and all the hydroxyl groups are involved in hydrogen bonding. This hydrogen bonding was proposed by Blackwell *et al.* and has since been adopted as the *de facto* structure.

β-Chitin is found in chitin extracted from the diatom spines and squid pens. Unlike α-chitin, but as with cellulose, β-chitin is packed in a parallel arrangement that does not favor the inter-chain hydrogen bonding between the C-6 hydroxyl groups along the c-axis, otherwise present in α-chitin [Figure 6.2]. This bonding arrangement facilitates the incorporation of water molecules between the weakly interacting β-chitin chains as would happen to β-chitin derived from diatom spines. In the case of squid pen β-chitin, there is complete dissolution in water.[2] It should be noted that because of the greater mobility of β-chitin chains, they are readily transformed irreversibly by steam annealing into α-chitin. In addition, the widespread availability of β-chitin is questionable to be of much utility in isolating and realizing chitin in a hydrophilic form at the present time.

Little is known of the solid-state structure of γ-chitin except that it is a mixture of α and β-chitin, with two parallel chains and for every anti-parallel stack that leads to water swelling properties intermediate between α and β-chitin.

The main consequence of the strong hydrogen bonding in α-chitin is its insolubility in all common organic solvents as well as aqueous acids making α-chitin a difficult polymer to work with. This intractable character, so essential for its biological role in the animal and plant kingdom, becomes a big hindrance for studying and utilizing α-chitin. It is a clear testimony to the tenacity of the early scientists in chitin research that the intractability of α-chitin has been overcome.

This insolubility property was so unique that it led to several patents being filed for

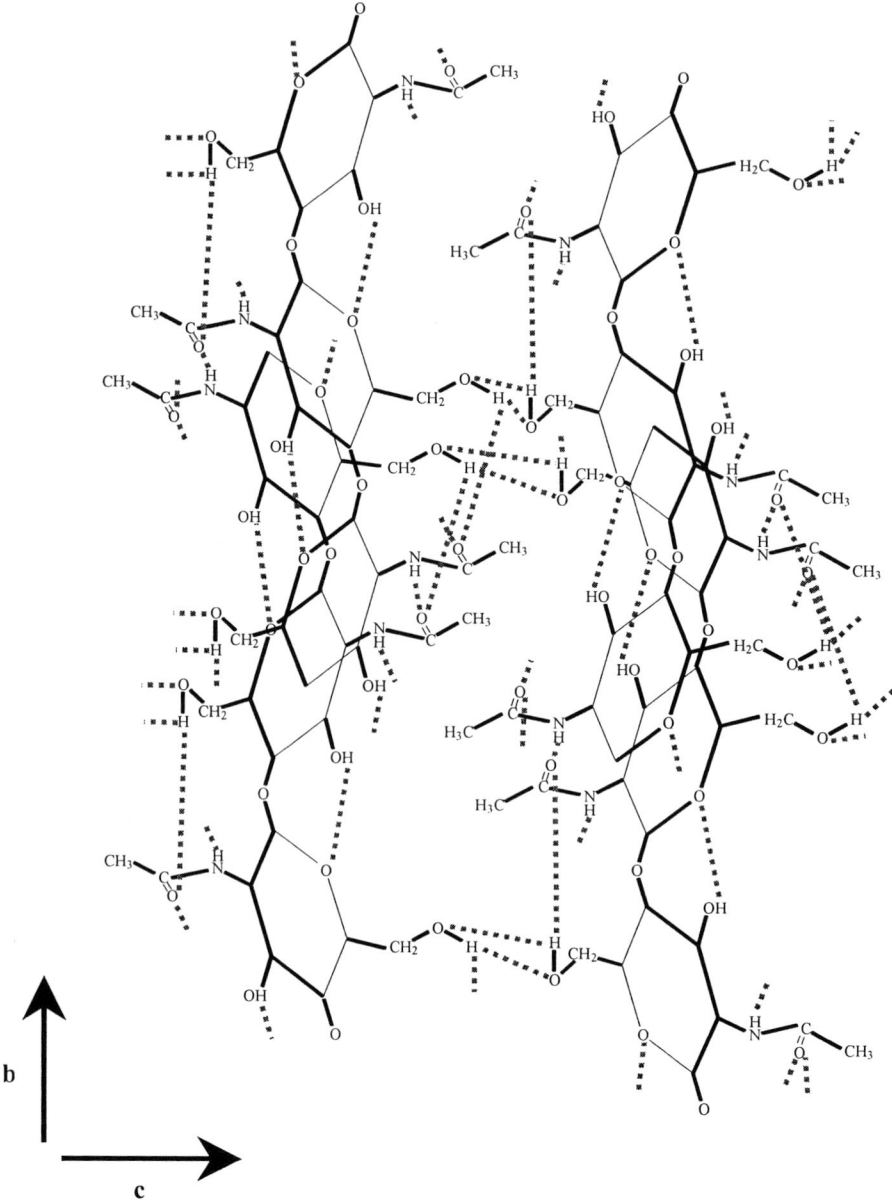

Figure 6.1:   The extensive H-bonding found in α-chitin[1]

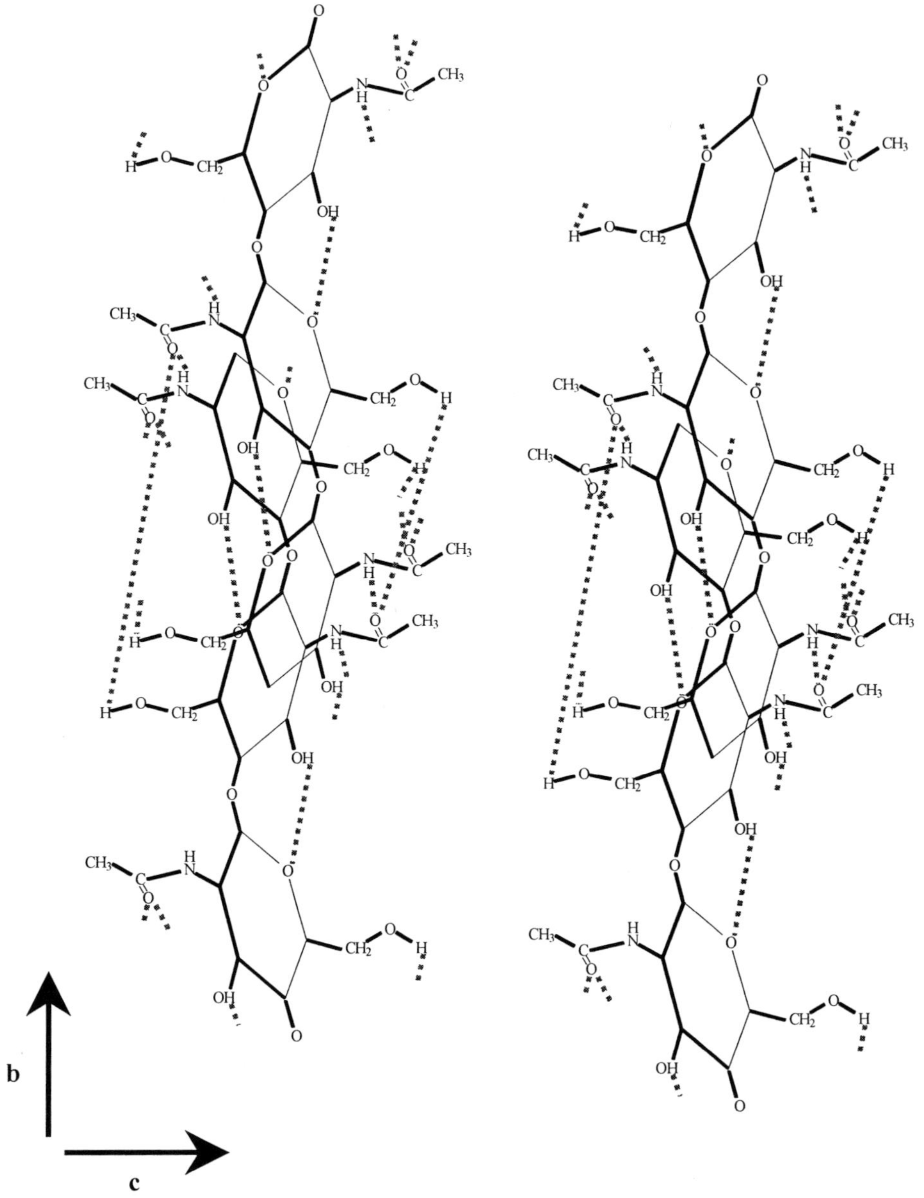

Figure 6.2:   The H-bonding found in β-chitin[1]

dissolving chitin in the early years. Austin first patented the use of cholorethanol-mineral acid mixtures to purify and regenerate chitin.[3] This later lead to the use of chloroacetic acids and finally to the now *de facto* solvent system of 5% lithium chloride (LiCl) in N, N-dimethylacetamide (DMAc) and the less used N-methyl-2-pyrrolidone (NMP) containing 5-8% LiCl revolutionizing the utilization of chitin.[4]

The development of chitosan resulted in the weaker hydrogen bonded network in the biopolymer. Coupled with the ready ability of the amino group to be protonated by aqueous dilute acids rendering a pathway for manipulating the biopolymer at pH <6. In fact, a casual survey of the earlier scientific literature reveals that more reports have been published on the chemical reactions and applications of chitosan. While chitosan has managed a large following, the potential to use chitin has just begun to be realized. It is the author's opinion that chitin will blossom into the biopolymer of the 21st Century.

6.3 CHARACTERIZATION OF CHITIN PROPERTIES

In the utilization of chitin, one major issue to-date has been the lack of consistency in the properties of the biopolymer and the methods used to determine these properties. As stated in the definition in Chapter 1, chitin and chitosan is found (after the isolation process) as a copolymer of N-glucosamine and N-acetyl-glucosamine. Therefore, the two most important determinants of chitin's structural properties are the degree of acetylation (DA) or deacetylation (DD) and molecular weight. The degree of acetylation (or deacetylation) identifies the biopolymer as chitin or chitosan, whereas the molecular weight determines the viscosity and rate of degradation. The residual protein, moisture content, ash content, lipid content, heavy metal content, impurity content and color are other properties most frequently reported. The characterization methods used to determine these properties also span a wide range, from economic methods for routine quality control to expensive instrumentation for obtaining sophisticated specialized information. It must be kept in mind that just like other polymers, the properties of chitin and chitosan are an average of the contribution of all chains in the polymer sample. Inhomogeneity arising from the varied sources, however, can complicate the summation. The aim of having properly defined characterization methods for chitin and chitosan is to have a means to report reliably, the bulk biopolymer behavior and properties. This is necessary to set the appropriate and consistent expectations of the biopolymer properties.

Table 6.1 summarizes the established characterization methods for chitin and chitosan listed in the chitin handbook.[5] The molecular weight of chitin and chitosan can be determined by viscometry, light scattering and gel permeation chromatography, bearing in mind that different solvents are required for chitin and chitosan. The degree of deacetylation and N-acetylation can be obtained by first derivative ultraviolet spectrophotometry, infrared spectrophotometry, enzymatic and chromatographic methods and by nuclear magnetic resonance (NMR). Finally, the NMR instrument can also be used to probe molecular structure. These methods are described in enough detail to be a good starting point to gain an appreciation for chitin characterization methods and the reader is referred it. It should be noted that the use of these methods to determine the various characteristics of chitin are for research and general utility. The results obtained are not refined in detail to the level that is required for submission to regulatory agencies, that includes performing the studies according to Standard procedures under good laboratory procedures (GLP), to be elaborated later in this book.

Property	Characterization method
Molecular weight	Viscometry Light scattering High performance gel permeation chromatography
Degree of acetylation	First derivative ultraviolet spectrophotometry Infrared spectrophotometry Enzymatic determination Chromatographic determination NMR (^{13}C, ^{1}H, liquid-state and solid-state)
Degree of N-acetylation	Metachromatic titration Dye absorption Infrared spectroscopy
Molecular structure	NMR (^{13}C, ^{1}H, liquid-state and solid-state)

Table 6.1: Established characterization methods for chitin and chitosan from *Chitin Handbook*[5]

Described below are discussions on three of the properties employing methods stated above to highlight the importance of that property and the method used to determine them.

6.3.1 Degree of N-Acetylation (DA)/Deacetylation (DD)

The DA is the number of glucopyranose units of the biopolymer chain having the N-acetyl groups attached to it. The DA is a key property that must be determined as it influences the physical and chemical properties of chitin and chitosan such as solubility, chemical reactivity and biodegradability and therefore their applications. An accurate and rapid method suitable for routine use is desirable.

Many methods have been developed to determine the DA including titration, infrared spectroscopy, high performance liquid chromatography (HPLC), circular dichroism, ultra-violet spectroscopy, near infrared spectroscopy, pyrolysis-gas chromatography, thermal analysis and nuclear magnetic resonance spectroscopy[6, 7, 8, 9, 10, 11]. Each method has their own advantages and limitations as it relates to concerns such as expediency, sample solubility requirement, cost and availability of instrumentation.

The most fundamental and well-utilized method is elemental analysis. The elemental ratio between carbon, hydrogen and nitrogen (CHN) obtained experimentally is used to fit the calculated CHN values based on their percentages to give the DA. This is somewhat crude and has to take into account bound water but for the early workers was an effective method not only to differentiate between chitin and chitosan but also in determining the degree of substitution for various reactions.[12] This method still has utility today not only as a first approximation method but can be quite accurate for determining the degrees of substitution when atoms such as phosphorus and sulfur are involved as there is a direct correlation with the nitrogen atom on the monomer. For this method to be as accurate as possible, exclusion of moisture and the best purity of material is crucial.

Perhaps the most discussed method in determining the DA and DD used considerably in research is infrared (IR) spectroscopy based on tracing the oscillation band of the amide I band of chitin.[13] This is because IR spectroscopy is relatively rapid, is a common instrument found in most research laboratories, sample purity is not as critical and the method can be used with insoluble samples using the KBr disc method. This renders the IR method the greatest utility over other methods, which require elaborate and time-consuming sample and other reagents preparation.

The IR spectrum of chitin in Figure 6.2 shows characteristic bands at absorptions approximately 1655, 1630 and 1560cm^{-1}, the first two have been assigned as the amide I bands while the latter the amide II band of chitin. The N-acetyl content is based on the infrared absorption (A) of one of these amide bands as a ratio with an internal absorption standard band. The amide band is chosen from either the amide Ia and Ib bands at 1655 and 1630cm^{-1} respectively or the amide II band at 1560cm^{-1}. The internal reference is selected from the absorption bands of the hydroxyl group at 3450cm^{-1}, the C-H stretching band at 2878cm^{-1}, the bridging oxygen-stretching band at 1160cm^{-1}, the C-O stretching bands at 1070, 1030 or 897cm^{-1}.[14] Several IR absorption band ratios are possible to determine the % of DA or DD. The amide Ia band at 1655cm^{-1} is preferred over the amide II band for analyzing samples of low N-acetyl content.[15] This is because at low N-acetylation, the -NH$_2$ band centered at 1590cm^{-1} predominates and obscures the amide II band, making its estimation difficult. The 3450cm^{-1} band is prominent and relatively isolated. The 2878 cm^{-1} band is dependent on the level of N-acetylation as it originates from the C-H band vibrations and therefore requires further calibration with other techniques. To date, the A_{1655}/A_{3450} ratio is the preferred ratio as it gives % N-acetylation in good agreement with the titrimetric method over a wide range.

The primary reservation of using the A_{1655}/A_{3450} band ratio is in the choice of the baseline (line A in Figure 6.3) for the amide I band. For a fully deacetylated material, this baseline gives a positive % N-acetylation value of between 3-5% even though the amide I band should essentially be absent.[16] However, for samples having a % N-acetylation greater than 20% the estimated DA or DD matches those values obtained using other methods, demonstrating the utility of this ratio for such samples.

An adjustment of the baseline was proposed (line B in Figure 6.3) by Miya et al to provide results of good accuracy for samples having DA less than 10% while retaining the advantages of the A_{1655}/A_{3450} ratio method.[17] Using this baseline, the equation for the calculation of DA is:

$$\% \text{ N-acetylation} = (A_{1655}/A_{3450}) \times 115$$

This revision has shown good agreement with the dye method over a wide range of % N-acetylation and is used routinely. A Standard protocol prepared in the Author's laboratory is included for reference as Appendix 2.

While IR is of great utility for determining DA, the first derivative uv-vis spectrophotometry

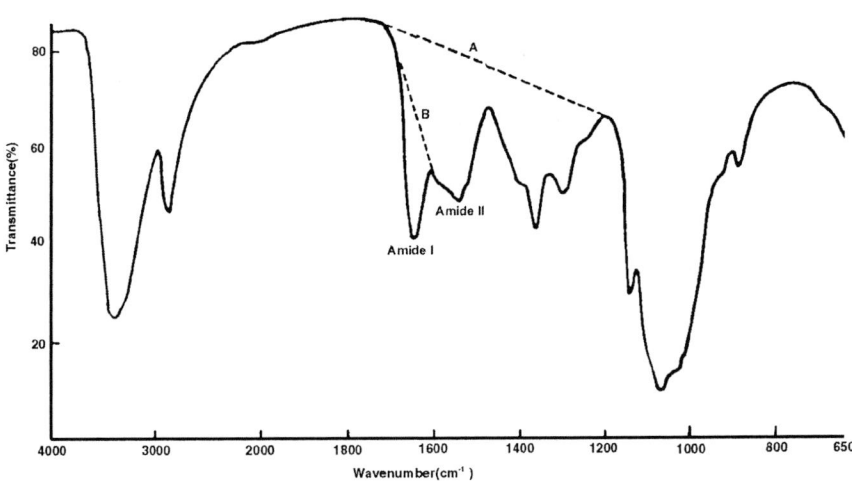

Figure 6.3:   The baseline assignment of the amide I absorbance band of chitin for the A_{1655}/A_{3450} ratio. The A dotted line is the baseline used originally and the B dotted line is the revised baseline

(Reprinted with permission form Elsevier Science Ltd)

method is being advocated as a simple, rapid and reliable method for determining the DD for chitosan on a routine basis especially for commercial quality control.[5] This is because of the ready solubility of chitosan and the simplicity of the uv-vis method. A Standard protocol prepared in the Author's laboratory is included for reference as Appendix 3. The nmr method is probably the most accurate, but instrumentation is expensive.

6.3.2 Molecular Weight Determination

The molecular weight of a polymer is the average of all the molecular weight of individual chains. The common methods of expressing polymer molecular weights are the number-average molecular weight (M_n), the weight-average molecular weight (M_w) and the viscosity-average molecular weight (M_v) and are normally dependent on their method of measurement. Most standard Polymer Chemistry texts have a discussion of each of these methods. The polydispersity index [P.I.] is given by the ratio of M_w/M_n. Normal synthetic polymers have a narrow P.I. of below 5. Proteins are typically the only biopolymers that are monodisperse. Most of the other natural occurring polymers such as cellulose and chitin have a high P.I. that can exceed 10.

The molecular weight of chitin and chitosan has been determined by a number of methods including light scattering spectrometry (LLS) that gives an absolute weight-average molecular weights (M_w). Gel permeation chromatography (GPC) and dilute solution viscometry are relative methods that require calibration based on an external standard to give the molecular weights.[18] The quality of chitin materials, the solvent system, temperature of analysis and viscosity of samples all play a role in obtaining good molecular weight results.

Most of the reports of molecular weights studies have been with chitosans as the ready solubility in dilute acids/acetate buffer systems renders chitosan to solution methods, the basis for all polymer molecular weight determinations. Muzzarelli et al have demonstrated the utility of laser light scattering spectrometry to a series of chitosans where the M_w values span a range of 1.9×10^5 to 7.0×10^5 for chitosan solutions in acetic acid/sodium acetate solvent system. In viscometry, the Mark-Houwink constants a and Km are important parameters that has been established by several workers. With these values, the M_v for chitosan solutions are readily obtained. The determination of molecular weight of chitosan by GPC was first reported by Wu and Bough and refined by Wu using dextran standards and aqueous acetic acid solvent system.[19] The M_w molecular weight as high as 2×10^6 were reported. Other workers have utilized pullulan as a reference for the molecular weight ranges and used sodium or ammonium acetate as a buffer with the acetic acid solvent system. The combination of GPC with LLS is also now popular.[20]

The limited choice of solvents and the nature of the solutions formed necessarily means that chitosan has been more studied than chitin. Nevertheless the molecular weight of chitin has been elucidated.[21] Therefore, the methods to determine the molecular weights of chitin is now well established and provide reasonable values for quality control as well as research purposes.

6.4 OUTCOME

In establishing the methods to identify the characteristics of chitin and chitosan reliably, the chitin scientific community has performed admirably. The quality control of chitin production can now be readily substantiated as the methods required in determining properties such as the degree of acetylation and molecular weight are now well documented, refined and easily available. Therefore, a raw materials producer, manufacturer or end-user can list the desired properties sought and can verify them readily.

The focus on IR does not imply its superiority over other methods in obtaining the DA or DD values of chitin and chitosan. However, it is the optimum method where profitability and feasibility are concerned, in terms of, simplicity, cost, up-scaling and reliable results. In characterization, it is desirable to choose a method that is suited for both chitin and chitosan. Not only will this avoid two separate assemblies of equipment operating on different parameters, but will also provide a more credible ground for comparison.

However, the choice of characterization technique in obtaining the DA or DD values will ultimately depend on the manufacturer's preference and the purity issue. Since sample purity is not critical in IR, other methods that can concurrently conduct both DA and purity measurements of the chitin material may be more expedient. For any particular manufacturer or researcher, it is important to note that one chosen characterization method applied across the board for all chitin-based products will inject consistency in product evaluation, a requirement in the development of standard protocols.

6.5 REFERENCES

[1]   R. Minke, J. Blackwell, The structure of α-chitin. J. Molecular Biology, 120 (1978) 167-181

[2]   S. Salmon, S.M. Hudson, Crystal morphology, biosynthesis and physical assembly of cellulose, chitin and chitosan. J.M.S.-Review. Macromolecular Chemistry and Physics, C37(2) (1997) 199-276

[3]   P.R. Austin, Purification of chitin. U.S. Patent: 3,879,377, Apr 22, 1975; P.R. Austin, Solvents for and purification of chitin. U.S. Patent: 3,892,731, Jul 1, 1975; P.R. Austin, Chitin solution. U.S. Patent: 4,059,457, Nov 22, 1977

[4]   P.R. Austin, Solvents For And Purification Of Chitin. U.S. Patent: 4,062,921, Dec 13, 1977

[5]   Various authors. in Chitin Handbook, R.A.A. Muzzarelli, M.G. Peters, eds., Atec Edizioni, Italy 1997. 87-143

[6]   E.R. Hayes, D.H. Davies, Characterization of chitosan: The determination of the degree of actylation of chitosan and chitin. in Proceedings of the First International Conference on Chitin/Chitosan, R.A.A. Muzzarelli, E.R. Pariser, eds., MIT Sea Grant Program, Cambridge, MA, USA, 1978. 406-420

[7]   F. Nanjo, R. Katsumi, K. Sakai, Enzymatic method for determination of the degree of deactylation of chitosan. Analytical Biochemistry 193 (1991) 164-167; F. Niola, N. Basora, E. Chornet, P.F. Vidal, A rapid method for the determination of the degree of N-acetylation of chitin-chitosan samples by acid hydrolysis and HPLC. Carbohydrate Research 238 (1993) 1-9

[8]   RAA Muzzarelli, R Rochetti, Determination of the degree of acetylation of chitosan by first derivative ultraviolet spectrophotometry. Carbohydrate Polymers 5 (1985) 461-472; SC Tan, E. Khor, TK Tan, SM Wong, The degree of deacetylation of chitosan: Advocating the first derivative UV-spectrophotometry method of determination. Talanta 45 (1998) 713-719

[9]   T.D. Rathke, S.M. Hudson, Determination of the degree of N-acetylation in chitin and chitosan as well as their monomer sugar ratios by near infrared spectroscopy. J. Polymer Science: Part A: Polymer Chemistry 31 (1993) 749-753

[10]   G. Alonso, C. Peniche-Covas, J.M. Nieto, Determination of the degree of acetylation of chitin and chitosan by thermal analysis. J. Thermal Analysis 28 (1983) 189-193

[11]   A. Hirai, H. Odani, A. Nakajima, Determination of the degree of deacetylation of chitosan by ^{1}H NMR spectroscopy. Polymer Bulletin 26 (1991) 87-94; K.M. Vårum, M.W. Anthonsen, H. Grasdalen, O. Smidsrød, Determination of the degree of N-acetylation and the distribution of N-acetyl groups in partially N-deacetylated chitins (chitosans) by high field nmr spectroscopy. Carbohydrate Research 211 (1991) 17-23

[12]   S. Hirano, Y. Ohe, H. Ono, Selective N-acylation of chitosan. Carbohydrate Polymers 47 (1976) 315-320

[13]   G.K. Moore, G.A.F. Roberts, Studies on the acetylation of chitosan. in Proceedings of the First International Conference on Chitin/Chitosan, R.A.A. Muzzarelli, E.R. Pariser, eds., MIT Sea Grant Program, Cambridge, MA, USA, 1978. 421-429; T. Sannan, K. Kurita, K. Ogura, Y. Iwakara, Studies on chitin: 7. I.r. spectroscopic determination of degree of deacetylation. Polymer 19 (1978) 458-459; G.K. Moore, G.A.F. Roberts, Determination of the degree of N-acetylation of chitosan. International J. Biological Macromolecules 2 (1980) 115-116; K.M. Vårum, B. Egelandsdal, M.R. Ellekjær,

Characterization of partially N-acetylated chitosans by near infra-red spectroscopy. Carbohydrate Polymers 28 (1995) 187-193

[14] Y. Shigemasa, H. Matsuura, H. Sashiwa, H. Saimoto, Evaluation of different absorbance ratios from infrared spectroscopy for analyzing the degree of deacetylation in chitin. International J. Biological Macromolecules 18 (1996) 237-242

[15] A. Domard, M. Rinaudo, Preparation and characterization of fully deacetylated chitosan. International J. Biological Macromolecules 5 (1983) 49-52

[16] A. Domard, Determination of the N-acetyl content in chitosan samples by c.d. measurements. International J. Biological Macromolecules 9 (1987) 333-336

[17] M. Miya, R. Iwamoto, S. Yoshikawa, S. Mima, I.r. spectroscopic determination of CONH content in highly deacylated chitosan. International J. Biological Macromolecules 2 (1980) 323-324; A. Baxter, M. Dillon, K.D. A. Taylor, G.A.F. Roberts, Improved method for i.r. determination of the degree of N-acetylation of chitosan. International J. Biological Macromolecules 14 (1992) 166-169

[18] G.A.F. Roberts, J.G. Domszy, Determination of the viscometric constants for chitosan. International J. Biological Macromolecules 4 (1982) 374-377; R.A.A. Muzzarelli, C. Lough, M. Emanuelli, The molecular weight of chitosans studied by laser light scattering. Carbohydrate Research 164 (1987) 433-442; W. Wang, S. Bo, S. Li, W. Qin, Determination of the Mark-Houwink equation for chitosans with different degrees of deacetylation. International J. Biological Macromolecules 13 (1991) 281-285; R.G. Beri, J. Walker, E.T. Reese, J.E. Rollings, Characterization of chitosan via coupled size-exclusion chromatography and multiple-angle laser light-scattering technique. Carbohydrate Research 238 (1993) 11-26; M. Terbojevich, A. Cosani, B. Focher, E. Marsano, High-performance gel-permeation chromatography of chitosan samples. Carbohydrate Research, 250 (1993) 301-314; J.Z. Knaul, M.R. Kasaai, V.T. Bui, K.A.M. Creber, Characterization of deacetylated chitosan and chitosan molecular weight review. Canadian J. Chemistry 76 (1998) 1699-1706

[19] A.C.M. Wu, Determination of molecular-weight distribution of chitosan by high-performance liquid chromatography. Methods in Enzymology 161 (1988) 447-452

[20] I. Hall, D. Gillespie, K. Hammons, J. Li, Molecular eight and conformation of chitosan determined through SEC³ and standalone light scattering and viscometry techniques. Advances in Chitin Science, Vol 1, A. Domard, C. Jeuniaux, R. Muzzarelli, G. Roberts, eds., Jacques Andre Publisher, Lyon, France, 1996. 361-371

[21] M. Hasegawa, A. Isogai, F. Onabe, Molecular mass distribution of chitin and chitosan. Carbohydrate Research 262 (1994) 161-166

CHAPTER 7: THE CHEMISTRY OF CHITIN AS IS KNOWN TODAY

7.1 MOTIVATION

Why should chemistry be done on chitin? Noted earlier, the strong inter and intra molecular hydrogen bonding in α-chitin had literally rendered the biopolymer structure essentially impregnable to solvents or reagents that in the process saddled chitin with the unflattering reputation of being **intractable**. Chemical manipulation was seen as one route to overcoming the intractability of chitin to make chitin more accessible. It was envisaged that the resultant chitin-derivatives produced would have modified biopolymer properties, improving solubility or swellabilty and processability, but optimistically, still retaining much of chitin's character. The outcome was the myriad of new compounds spawning a multi-faceted perspective to, and extending the science and application potential of chitin. A further incentive was the controlled hydrolysis of the biopolymer to obtain lower molecular weight chitin and oligomers. These forms of chitin would be easier to manipulate and have applications yet to be conceived. These original aims to utilize chitin fully have relevance to biomedical applications, because one of the exercises for many new chitin derivatives was to evaluate their potential as biomedical products. Finally, biomaterials must be produced in a usable, pure and reproducible form that is largely dependant on the chemistry used to produce them.

What can be done to chitin? In chitin, the C-6, and C-3 positions of the monomer contain hydroxyl groups and in chitosan there is the additional N-2 amino functionality that can participate in chemical reaction. All three sites are available for chemical reaction and therefore, the chemistry of chitin and chitosan has been principally one of chemical derivatization of the functional groups. The C-6 site dominates reactivity, as being a pendant group from the pyranose ring is more exposed and readily accessible to incoming reagents. As the course of reaction progresses, the polymer chains should move further apart and the C-3 position becomes increasingly available for reaction. Most reactions of this nature give mixed results, with the C-6 position dominating the derivatization and some on the C-3. In most instances, the degree of substitution at the two sites is not reported especially for studies conducted prior to the introduction of more sophisticated and sensitive instrumental techniques. Together with the averaging character of polymer science, the degree of substitution has always been implied as on the C-6 position.

Historically, the intractability of chitin dictated heterogeneous chemical reactions as the starting point for scientists of the day to commence unraveling the chemistry of chitin. Heterogeneous chemical reactions as a rule are difficult to define and regulate. This implies that optimizing reaction conditions to obtain a high degree of substitution and distinct chemical derivatives is demanding. Therefore, subsequent to heterogeneous chemical reactions, finding better and more efficient methods of reacting or modifying α-chitin became a priority. Over the years, this led to improvements in reaction conditions of the heterogeneous reactions, the introduction of tosyl, trityl and silyl groups and other intermediates to facilitate better chemical modifications; the use of β-chitin as an alternative to α-chitin and the introduction of post-shape chemical modification procedures. Concurrently, homogeneous reactions were conducted beginning with strong acids culminating in the introduction of homogeneous reactions with the chitin solvent 5%LiCl/DMAc. All this effort has led to a better understanding of the chemical modification reactions of chitin. Detailed below is an overview, outlining the varied chemical history of

84

chitin. Coupled with the examples reported in the biomedical applications survey, the wide range of chitin derivatives provides the backdrop for appreciating the appropriate perspective to consider chitin for biomedical applications.

7.2 ALKALI-CHITIN

The first and for a long time, the *de facto* method to prime chitin to be chemically responsive has been to use alkali i.e. sodium hydroxide (NaOH). This method is largely an extension of the "alkali-cellulose" process based on cellulose chemistry. Alkali-chitin is essentially the conversion of the C-6 hydroxyl of the sugar ring into an "alkoxide", the anion form together with the associated sodium counter-ion (Figure 7.1). The preparation involves exposing chitin to an aqueous-based strong alkali medium in several soak, impregnate and rinse cycles. The eventual extensive swelling gives alkali-chitin, succumbing chitin to some form of processability. In this form, alkali-chitin is reacted heterogeneously with various chemical precursors of the derivatives introduced normally in a mixed solvent system, to form

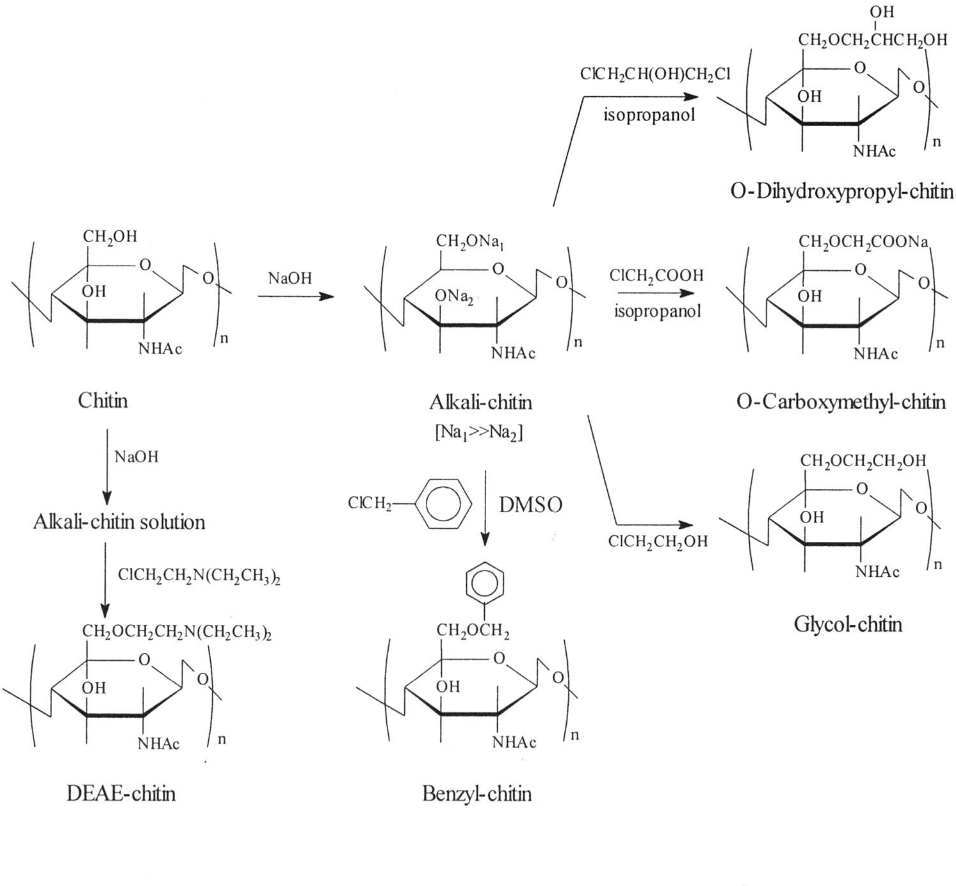

Figure 7.1   Preparation and reactions of Alkali-chitin

derivatives of chitin. The extent of reaction is a function of the reactivity, solubility in aqueous/organic solvent mixtures and potential steric factors of the derivatizing reagent and the time of reaction.

The first use of NaOH to prepare alkali-chitin is ascribed to Thor, covered in patents granted in 1939/40.[1] Thor showed that a sodium ion incorporation of up to 90% could be achieved with 50% NaOH solution. The high amount of sodium hydroxide involved in the alkali-chitin process creates a very harsh environment that frequently degrades the chitin polymer chains. In addition, N-deacetylation, the removal of the acetyl group from the N-2 position also occurs that can lead to the formation of amphoteric polymers having both carboxyl and amino groups. For example, Sannan et al has shown that alkali-chitin films left at 25°C over 2 days exhibit detectable deacetylation. The chitin became water soluble with accompanying chain degradation and destruction of chitin's secondary structure.[2] Low temperatures, preferably below freezing, arrests these side effects and the use of freezing temperatures is now a common practice when working with alkali-chitin.

Alkali-chitin is now, by and large, prepared by dispersing chitin powder in NaOH and allowed to stand for 1 to 3 hours at room temperature. Subsequently, the alkali-chitin is poured onto crushed ice that after filtering gives a clear alkali-chitin solution. Alternatively, chitin is subjected to two or three cycles of freezing and thawing with removal of aqueous liquid to give alkali-chitin for subsequent reaction. The concentration of NaOH used in both procedures spans from 40 to 50%. Chitin derivatives have been generated from both alkali-chitin producing sequences.

The alkoxide ion is a very strong base and readily reacts with substituted alkyl and aryl halides to form the respective ether based derivative. One of the most established examples of the use of alkali-chitin is the carboxymethylation of chitin that proceeds with monochloroacetic acid in isopropanol performed at 0°C (Figure 7.1).[3] Depending on the duration of exposure to NaOH, degrees of O-carboxymethlylation ranging from 0.1 to 1.0 have been reported. Tokura et al have also prepared carboxymethyl-chitin (CM-chitin) and dihydroxylpropyl-chitin (DHP-chitin) using the alkali-chitin method, with degrees of carboxymethylation of 0.6 and 0.9 respectively.[4] A degree of substitution (d.s.) above 0.60 renders CM-chitin water-soluble.[5] In its sodium salt form, the CM-chitin is a good chelating agent as demonstrated by its high affinity for calcium ions.[6] Interestingly, when the carboxymethylation reaction is performed in the absence of isopropanol and high alkali content at 30°C, O-carboxymethlylated chitosan is obtained as the reaction conditions in this instance favors concurrent deacetylation.[7] In similar manner, ethylene chlorohydrin has also been reacted with alkali-chitin solution to prepare glycol-chitin.[8] Kurita et al have also used the alkali-chitin solution to produce C-6 diethylaminoethyl derivatives that have a d.s. as high as 1.4 indicating extension of reaction to the C-3 position.[9] Tokura et al have also demonstrated that it is possible to prepare benzyl-chitin from alkali-chitin obtaining a substitution of 0.75 as the best result.[10]

Efforts to circumvent and improve on the alkali-chitin method first led some researchers to start with the more readily manageable chitosan as an alterative route to furnish the desired chitin derivatives.[11] Hirano et al demonstrated the utility of this direction by producing N-acetyl-chitosan, essentially regenerated-chitin, as a precursor to form alkali-chitin solutions (Figure 7.2).[12] The sequence in the procedure starts with chitosan being dissolved in aqueous acetic acid, followed by diluting with methanol, reacting with acetic anhydride to give N-acetyl-chitosan. Swelling in NaOH, pouring into crushed ice to give the regenerated chitin-based form of aqueous alkali-chitin solutions. The utility of this chitosan derived alkali-chitin is demonstrated as with reaction with appropriate dialkyl-sulfates gives O-alkyl derivatives and in preparing ((sodiumthiol)thiocarbonyl)-chitin or chitin xanthates from carbon disulfide and other sulfur containing derivatives similar to cellulose-xanthate or viscose used in the manufacture of rayon.[13, 14] Naturally, N-acetyl-chitosan generates CM-chitin with little difficulty.[15] Further reaction of this CM-chitin leads to N-acyl O-CM-chitosans by first deacetylating CM-chitin under mild conditions using aqueous 10% NaOH at 90°C to give O-CM-chitosan. The O-CM-chitosan was next reacted with a series of carboxylic anhydrides, generating a series of N-acyl derivatives of O-CM-chitosan.

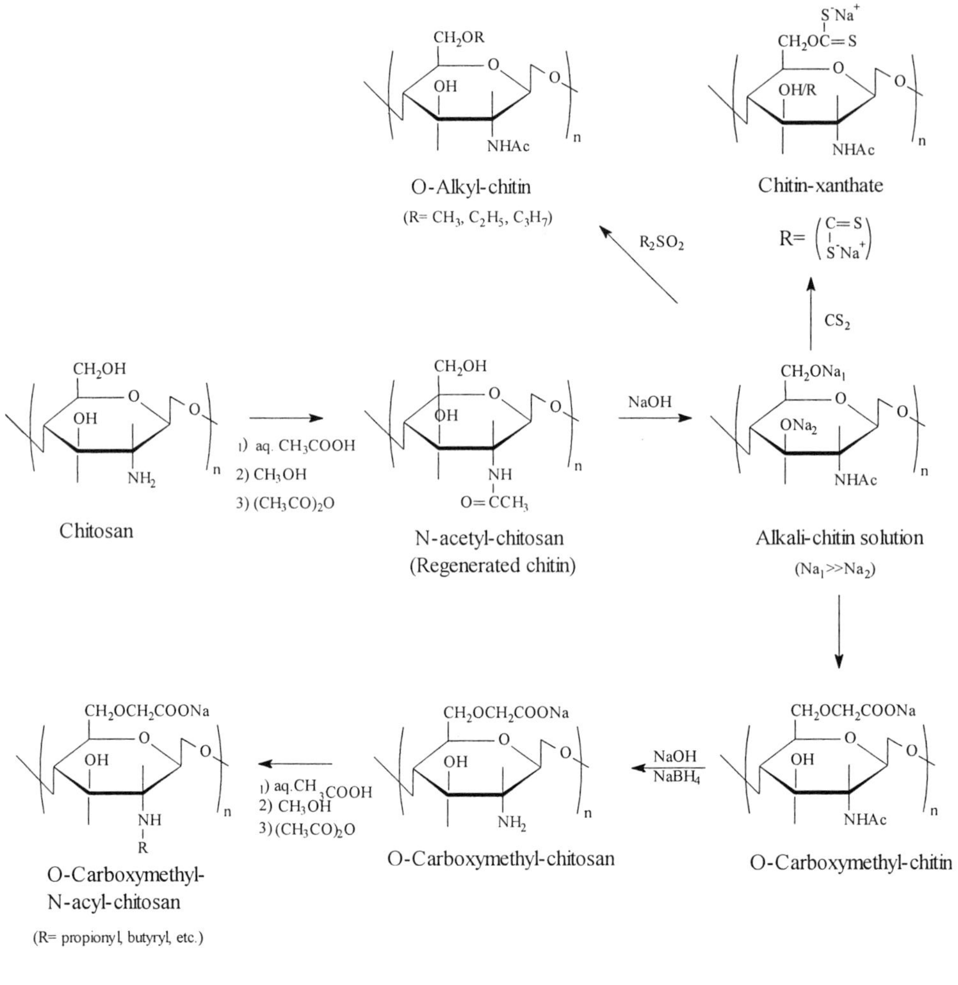

Figure 7.2   Alternate chitosan route to C-6 chitin derivatives

The utilization of alkali-chitin to obtain C-6 derivatives of chitin is limited, not only because of the inefficient use of NaOH, but the constraint of alkyl and aryl halide reagents means only a finite number of derivatives can be prepared. The aromatic-based halides are also more difficult to synthesize, as the bulky benzene ring impedes access to the C-6 site. The alternate chitosan route being less troublesome extends the list of chitin derivatives. However, the alkali-chitin reaction was an achievement in its own right that set the stage for the next wave of chemical innovations in chitin.

7.3 TOSYL-CHITIN

In a series of reports commencing in 1991, a refinement of the alkali chitin process to facilitate better control and reactivity was developed and largely credited to Kurita and co-workers. This has been achieved with a reaction called tosylation, to form a new intermediate, tosyl-chitin, more suitable for subsequent chemical reaction (Figure 7.3).[16]

In tosylation, the position occupied by the sodium ion on the C-6 hydroxyl group of chitin is replaced by a tosyl group, a sulfonated aromatic moiety, well known in organic chemistry as a good "leaving group". A leaving group is so called because its chemical role is as an intermediary. In tosylation, the tosyl group's chemical reactivity permits it to, first react with an otherwise difficult to react position and occupy it. Subsequently, the tosyl group is readily replaced by suitable chemical entities in chemical substitution reactions.

Tosyl-chitin was first synthesized by the interfacial reaction of aqueous alkali-chitin with a tosyl chloride-chloroform solution. Tosyl-chitin was found to be hydrophilic and swelled in water when the d.s. was below 0.3 but became hydrophobic and soluble in organic solvents when the d.s. was >0.4.[17] These results are clear indication of the tosyl group disrupting the hydrogen-bonding network for d.s. <0.3 and exhibiting its organic character for d.s. >0.4. The ready reactivity of tosyl-chitin with sodium iodide produced iodo-chitin at the C-6 position that can be reduced to the methyl group with NaBH$_4$ giving deoxy-chitin. Tosyl-chitin has also been used to prepare mercapto-chitin in a two-step procedure.[18] Iodo-chitin was found to be more soluble in organic solvents compared to tosyl-chitin while mercapto-chitin swelled in a number of organic solvents. Both iodo-chitin and mercapto-chitin have been shown to be good precursors for the graft copolymerization with styrene to produce chitin-polystyrene copolymer.[19]

The preparation of tosyl-chitin under homogeneous conditions using the LiCl/DMAc solvent system has also been reported.[20] It was found that O-acetylation was a predominant side reaction and only in the presence of a strong base was the reaction driven to favor the production of tosyl-chitin. The tosyl-chitin obtained under these conditions was subsequently shown to produce deoxy-thiocyanato-chitin in good yield.

Tosyl-chitin can be considered as the first attempt to breakaway from the heterogeneous reaction concept spawning reactive intermediates such as trityl-chitin and silyl-chitin that were not derived using aqueous alkali-chitin.

7.4 DIRECT REACTIONS

In chemical reactions, direct or homogeneous pathways are preferred as two or more components can interact more efficiently when mediated by the solvent. Therefore, with chitin, this situation would be ideal, as the biopolymer chains would be well dispersed in the

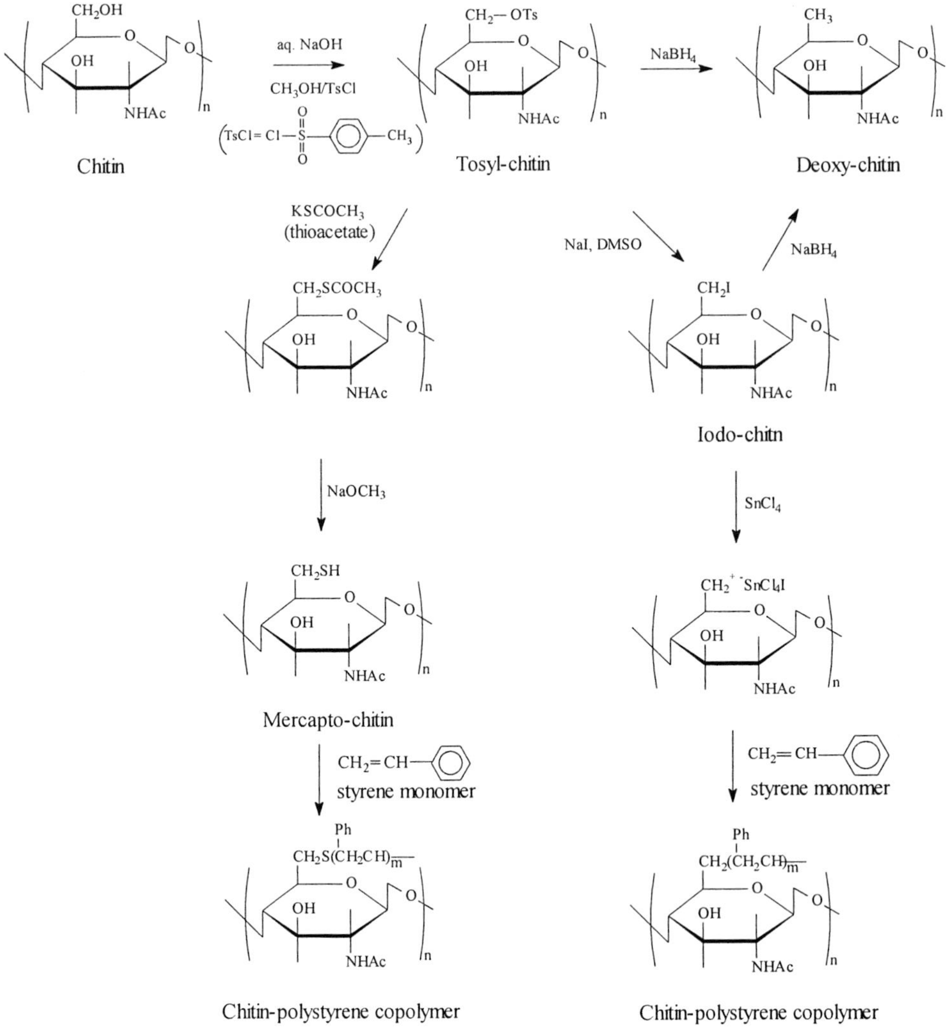

Figure 7.3   Preparation of Tosyl-chitin and its subsequent reactions

solvent exposing the C-6 hydroxyl group that is otherwise less accessible in the heterogeneous setting. Direct reactions became possible once solvent systems, mainly concentrated acids in the first generation, were established for chitin.

Some of the pioneering work can be credited to Tokura and co-workers with a series of studies with a methanesulfonic acid and glacial acetic acid mixture (Figure 7.4).[21] In a representative procedure, chitin powder was dispersed in the methanesulfonic acid and glacial acetic acid mixture, acetic anhydride included and stirred for 4 hours at 0°C and maintained overnight at that temperature. Upon precipitating the reaction mixture in cold water, diacetyl-chitin was recovered. Similarly, benzoyl-chitin (ref 4) and p-substituted benzoyl-chitins were prepared with this method that is a departure from the alkali method first used for the benzyl counterpart (Figure 7.1) and extended to prepare aliphatic substituents of formic, propionyl and butyryl chitins.[22, 23]

Highly phosphorylated-chitin at both the C-6 and C-3 positions, was prepared by reacting chitin with phosphorus pentoxide (P$_2$O$_5$) mediated in methanesulfonic acid that was water-soluble and found to be good metal chelators.[24] Inorganic esters of chitin have been demonstrated with nitric acid to give nitrates and sulfuric acid to give sulfates. Finally, 70% perchloric acid has also been used, in the reaction of chitin with butyryl chloride at 0°C to give butyryl-chitin on both the C-6 and C-3 positions.[25] This reaction, while heterogeneous in nature, is included in this section as the discussion is on the use of concentrated acids.

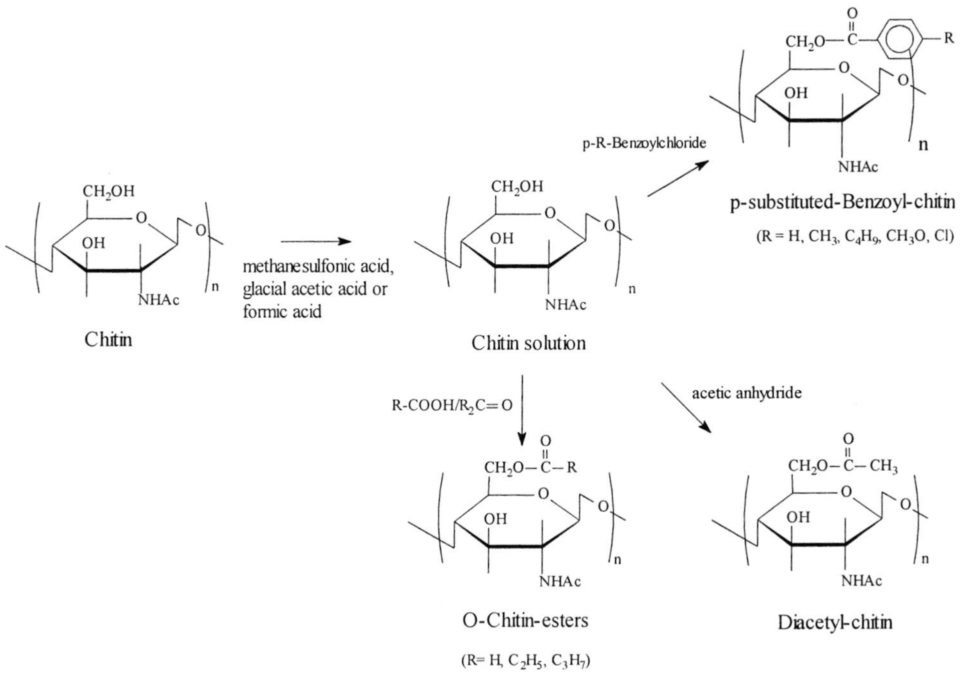

Figure 7.4   Direct reactions of chitin

It has to be noted that the reagents described here are harsh concentrated acids that require near ice temperatures to prevent depolymerization similar to the situation with alkali-chitin. Yet, chitin is degraded considerably under the influence of these strong acids and optimization is difficult to achieve. In a move away from such harsh reagents as well as to obviate the use of alkali-chitin, Kurita et al utilized a 50% acetylated chitin known to be water-soluble. This water-soluble chitin was reacted with a series of aldehydes to produce alkylidene substituted chitin obtaining reasonable degrees of substitution for hexanal, decanal and dodecanal substituents.[26] This can probably be noted as the bridge between the first generation of solvents based on acids to the true homogeneous reactions with the advent of the solvent system 5% LiCl/DMAc, first used to purify and study the properties of chitin.

Terbojevich et al. first showed that a chitin-5% LiCl/DMAc solution readily reacts with acid chlorides and anhydrides, isocyanates and SO_3-pyridine complex to give chitin-esters, -carbanilate and -sulfonate derivatives (Figure 7.5).[27] Vincendon extended the list with chitin carbamates by reacting chitin solution with aromatic isocyanates.[28] Simlarly, Morita et al have studied reactions under homogeneous conditions using 5% LiCl/DMAc with acetic anhydride to prepare O-acetyl chitin.[29] Homogeneous conditions were also exploited to react

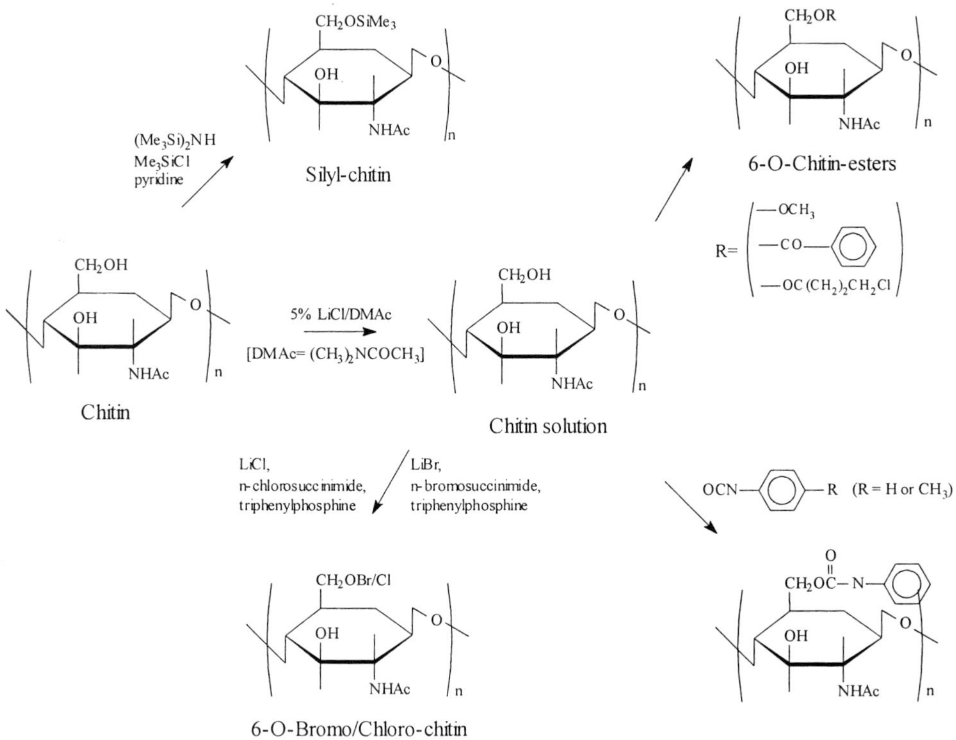

Figure 7.5   Homogeneous reactions of chitin

chitin with N-chlorosuccinimide and triphenylphosphine to give chlorodeoxy-chitin.[30] In the preparation of bromodeoxy-chitin, the corresponding bromine reagents LiBr with the N-bromosuccinimide-triphenylphosphine combination were utilized.[31] In yet a further variation, Tseng et al employed sulfuryl chloride to prepare chlorodeoxy-chitins.[32] Finally, Chow and Khor have prepared several C6-fluorinated-chitin derivatives with the chitin-5%LiCl/DMAc system and fluorine containing reagents.[33] From these examples, it would appear that this solvent system will become entrenched, conquering the hitherto intractability label of chitin.

To complete the chemistry picture, Kurita et al have recently reported the preparation of silylated-chitin as a new organosoluble precursor for facile modifications and film casting.[34] The free amino groups on chitin were first acetylated to exclude the possibility of N-silylation. This was followed by reaction of α-chitin and β-chitin with a mixture of hexamethyldisilazane and chlorotrimethylsilane in pyridine at 70 °C. High silylation degrees of 1.45 and 2.0 were obtained respectively.

In summary, starting with heterogeneous reactions, chitin chemistry has graduated to homogeneous reactions, first with harsh acids leading to the now standard 5% LiCl/DMAc solvent system. Acylation of chitin to generate organic ethers and amides have been demonstrated. Inorganic esters, halogenated compounds, carbamates, tosylation and silylation are springboards to a wider versatility to give rise to new derivatives. Chitin chemistry has therefore come a long way, the present methods capable of producing chitin derivatives, limited by the imagination of the individual scientist, that is a platform to revolutionize the utility of chitin.

7.5 β-CHITIN

The resourcefulness of chitin scientists has not been limited to α-chitin in the quest to exploit chitin. In the 1990's, β-chitin isolated from squid pens was seen as an alternative to address the intractability issue of α-chitin. The lower extent of hydrogen bonding in β-chitin permits easier access by common solvents facilitating homogeneous chemical reactions.[35] β-Chitin is readily extracted under milder conditions than α-chitin and is easily deacetylated to give chitosan. Some of the reactions that β-chitin have been subjected to include the methanol/acetic anhydride mixture in the presence of pyridine to give diacetyl/chitin and tritylation and tosylation by swelling β-chitin in pyridine and reacting with trityl chloride and tosyl chloride respectively to give trityl chitin and tosyl-chitin (Figure 7.6).[36] These facile reactions show that β-chitin is relatively versatile as a starting material for the chemical modification of chitin.

It should be noted that β-chitin is at present limited in supply and the longer-term utility of using this form of chitin in the preparation of chitin derivatives on a commercial scale may not be realized. It would also be instructive to compare the relationship of α-chitin to β-chitin. In fact, Kurita has shown that tritylation is higher and achieved in one-step with β-chitin whereas α-chitin requires a 5-step reaction sequence.[37] Further studies should answer questions such as comparing the properties of the same derivative from the two types of chitin.

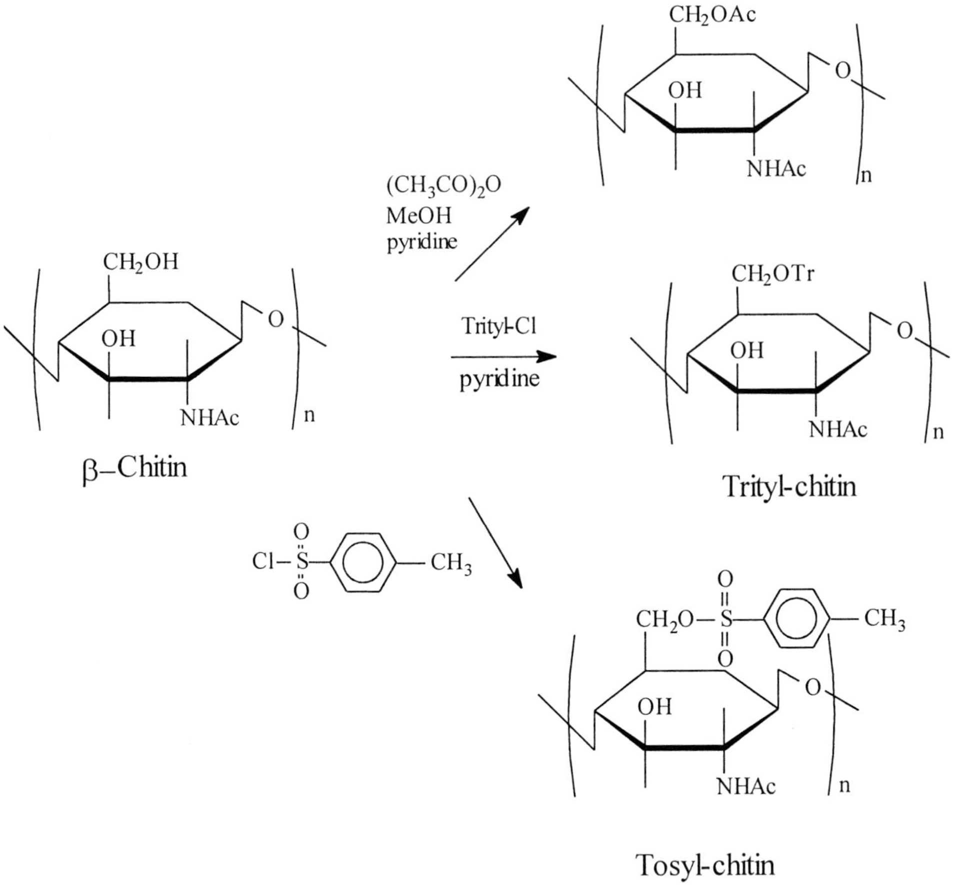

Figure 7.6   Reactions of β-chitin

7.6 CHITOSAN

The chitosan story, armed with ready solubility of chitosan in aqueous acids facilitating homogeneous chemical reactions, has been one that appears to have overshadowed chitin. One of the most established reactions of chitosan has been the conversion of the amino group to an acetyl group by N-acylation of chitosan with acetic anhydride in dilute acetic acid/methanol mixture to give regenerated chitin.[38] This straightforward approach has also been very successful in producing a wide range of chitosan derivatives from the respective carboxylic anhydrides to give the corresponding N-substituted alkyl and benzoyl-chitosans (Figure 7.7). The N-acyl-chitosans are however, gelatinous and insoluble in most solvents except for the C2 to C4 alkyl derivatives that were soluble in formic acid. Fujii et al noted the insolubility of these N-acyl derivatives and set about finding an alternative. By boiling chitosan in long alkyl-chain carboxyl-chlorides, specifically hexanoyl, dodecanoyl and tetradecanoyl, in chloroform/pyridine mixture, Fujii et al produced mixed N, O-acyl-chitosan derivatives that were soluble in chloroform, benzene, ether and pyridine after recovery.[39]

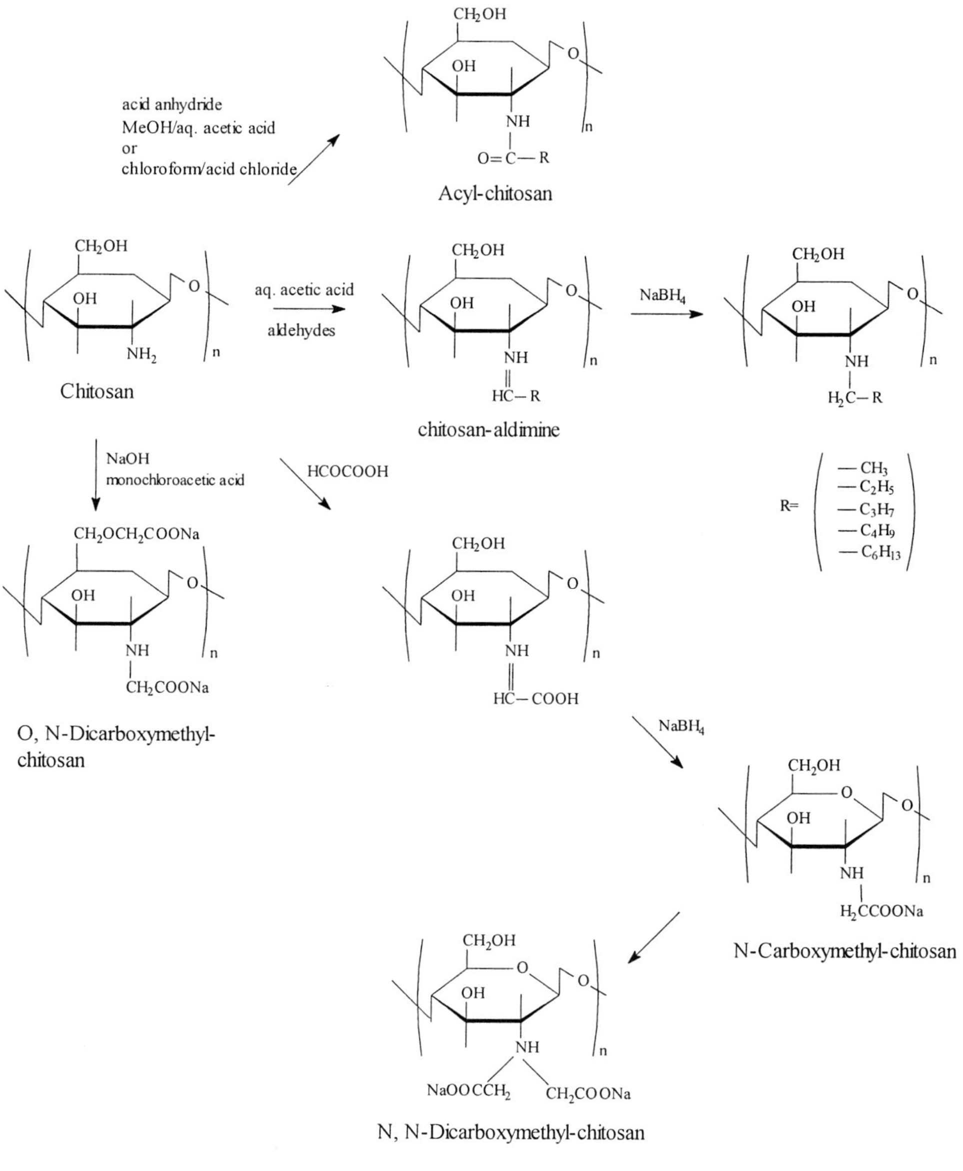

Figure 7.7   Reactions of chitosan

Therefore, it is clearly evident that while chitosan readily participates in homogeneous reactions, the utility of the derivatives depends on their resultant properties especially with regards to solubility. Several other examples of the reactions of chitosan include the reaction of chitosan dispersions with β-butyrolactone in dimethyl sulfoxide at 100°C to produce N-hydroxyacyl-chitosan that wasfound to be insoluble but could swell in polar solvents.[40] N-benzyl sulfonated-chitosans have also been prepared using the aldehyde group as the point of reaction with the amine functionality of chitosan.[41] A water soluble derivative of N-[(3'-hydroxy-2',3'-dicarboxy)-ethyl]-chitosan has also been synthesized.[42]

Apart from being a convenient homogeneous route to prepare CM-chitin, the preparation of CM-chitosan derivatives have also been investigated. Muzzarelli et al demonstrated that N-carboxymethylation could be obtained by first reacting the amino group on chitosan with glyoxylic acid to give the intermediate aldimine. Subsequent reduction gives N-CM-chitosan that is readily soluble in water for the whole pH range (Figure 7.7).[43] In subsequent work, it was demonstrated that excess glyoxylic acid would result in a double carboxymethylation at the nitrogen atom. The solubility and structure determination showed that there was N-mono- and N, N-di-carboxymethylated-chitosans.[44] Mixed N, O-CM-chitosans can also be obtained by reacting chitosan suspended in 42% NaOH with monochloroacetate to give the O-CM-chitosan that is subsequently reacted with glyoxylic acid and sodium borohydride successively to give the N, O-CM-chitosan. The distribution of CM-substituents has been found to be the C6 preferred over the C3 for N, O-CM-chitosan as expected and a 70/30 ratio of mono- and di-substituted N-CM-chitosan.[45]

Kurita et al developed facile reactions with N-phthaloylation as a convenient precursor to produce derivatives by reaction of chitosan with phthaloyl chloride (Figure 7.8).[46] 6-O-tritylation followed by 3-O-acetylation and detritylation with dichloroacetic acid gave 3-O-acetyl, N-phthaloyl-chitosan. Sulfation and chain branching was also demonstrated in the series of reactions developed. β-Chitosan was found to be more reactive compared to its α-chitosan counterpart.[47] The trityl-chitin/chitosan intermediate has also been used to develop chitin and chitosan-sulfates that have been investigated for their anti-HIV activity.[48]

Preparation of macromolecular complexes (MC) with chitosan has been demonstrated between glycol chitosan or methyl glycol chitosan and potassium metaphosphate.[49] The MCs were found to be insoluble and composed of tight networks when calcium ions were present to crosslink the network. Other MCs that have been produced include the interaction of chitosan with glycosaminoglycans (GAG) chondroitin sulfate and hyaluronate based on electrostatic interactions between the protonated amine ($-NH^{3+}$) and the anionic sulfate and/or carboxylic groups ($-OSO^{3-}$, $-COO^-$) on the GAGs.[50]

The outline shows the versatility and possibilities of homogeneous chitosan reactions with the only reservation being resultant derivatives may not be readily soluble.

7.7 FIBERS AND GELS

Chemical derivatization has been a channel in revealing the rich potential of chitin and chitosan. Despite this, it must be remembered that the base chitin has tremendous worth in its own right and for biomedical applications, the less done the better for a material. Researchers have also addressed the challenge of the intractability of chitin by exploring

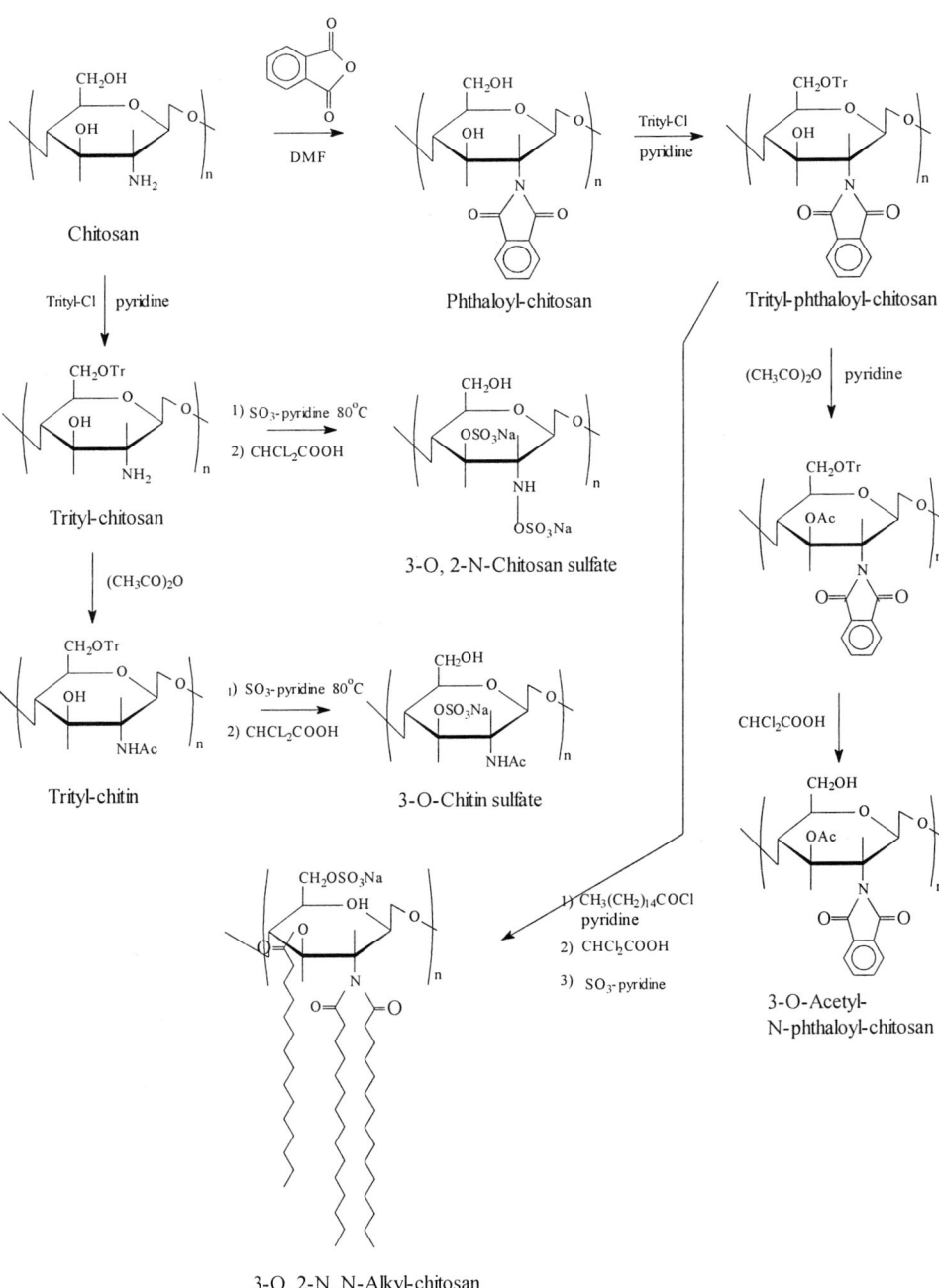

Figure 7.8   N-Phthaloyl reactions of chitosan

other approaches to prepare chitin without changing its original character. The extensive accounts in preparing chitin fibers, gels/hydrogels, beads and films are testimony to this.

In the making of chitin fiber, chitin solution is first prepared and subjected to the wet-spinning process, essentially pouring the chitin solution through small holes called spinnerets extruding long thin strands that coagulate in a bath containing a non-solvent to yield the fiber. The first chitin fibers produced used concentrated sulfuric acid as solvent that was probably difficult to handle and may not have been very successful.[51] More in line with the rayon process in cellulose fiber prodcution, chitin fiber was subsequently prepared from alkali-chitin and with chitin xanthate including mixing with cellulose xanthate to produce better quality fibers. Tokura et al prepared chitin fibers by soaking chitin powder in 99% formic acid at -20°C for 1 day, warmed to room temperature. The soak/warming process was repeated several times until a clear chitin gel was obtained.[52] This clear chitin gel was dispersed in dichloroacetic acid and isopropyl ether and spun into fine chitin fibers. Acetylchitin fibers were also similarly prepared.[53] With the advent of the LiCl/DMAc solvent system, it was natural to expect the production of chitin fibers using this system. The LiCl content influenced the viscosity that in turn affected the final quality of the fiber.

The preparation of chitosan fibers has surprisingly lagged behind chitin despite its ready solubility in aqueous acetic acid.[54] The first noted report surfaced in 1980 where a 3% chitosan solution was extruded into a coagulation bath containing aqueous NaOH that in some instances had copper salts added, possibly to enhance final fiber quality. Chitosan fibers are known to have reduced tensile strength when wet attributed to the hydrophilic nature of chitosan. Hudson et al crosslinked chitosan fibers with epichlorohydrin via the amino groups, improving the wet strength.[55] Most recently, Hirano et al have used N-acetyl-chitosan and N-propionyl-chitosan to prepare chitosan fibers, and blends with cellulose. The corresponding N-acyl-chitosan solutions were first prepared with alkali and added to crushed ice to give the alkaline solution that was subsequently reacted with carbon disulfide to give the final xanthate solution for wet spinning.[56] The N-priopionyl-chitosan-cellulose fibers had better mechanical properties compared to N-acetyl-chitosan-cellulose fibers. Hirano et al also produced chitosan fibers using aqueous acetic acid, post treating the fibers with a series of aldehydes without affecting fiber properties.[57] It is obvious that chitin and chitosan fibers, blends and even post-chemical treatment are all possible, demonstrating yet again that chitin is manageable for exploitation.

Gel forms is yet another variation to the chitin accessibility collection. The preparation of chitin gels can be traced back to Hirano et al who produced N-acyl, N-arylidene and N-alkylidene-chitosan gels using aqueous acetic acid solutions of chitosan, reported to be gelatinous and hygroscopic materials.[58] Moore and Roberts studied the reaction variables in detail and found that gelation was a consequence of the extent of N-acylation and determined that the mechanism was related to the concentration of chitosan and molecular weight of the acyl anhydride.[59] Hirano et al also prepared a thermoreversible chitosan gel based on oxalic acid.[60] Chitin gels obtained from solutions of LiCl/DMAc were first reported by Hirano et al to be rigid and transparent and readily hydrolyzed by hen-egg white lysozyme.[61] This approach was subsequently extended to N-acetyl and N-propionyl chitosan hydrogels from aqueous NaOH solutions combined with their respective cellulose composites from xanthate solutions.[62]

Formation of interpenetrating polymer network gels having a high affinity for unbound water based on aqueous acetic acid solutions of β-chitosan reacting with polyethylene-glycol and fabrication of a chitosan-cyclodextrin network by the reaction of oxidized polyaldehyde/β-cyclodextrin with the amino functionality of chitosan are known.[63, 64]

Figure 7.9   Chitin gels, fibers and beads (dehydrated) (clockwise from the top)

One recent method combining gel and film characteristics was the introduction of the concept of post shape chemical modification by Khor et al.[65] In a first example, a chitin hydrogel was converted into alkali-chitin on the surface and reacted with monochloroacetic acid to give a bilayered gel of surface caraboxymethyl-chitin and an inner core of chitin. The uniqueness of this procedure was that physical entanglement of the original chitin fibers in the hydrogel made the bilayered gel take up more water without dissolution. Subsequently, a chitin film was similarly treated in the solid state to give the bilayered feature that was found to have a reversible water-swellable gel character, since the surface i.e. the film could take on water to become a gel and dried back into film form reversibly.[66] This reaction was also extended to chitin beads.[67] Figure 7.9 shows representative samples of chitin gels, fibers and beads in the dehydrated form.

7.8 RANDOM OR BLOCK DISTRIBUTION OF N-ACETYL GROUPS DURING DEACETYLATION REACTIONS

The degree of acetylation or deacetylation has influence on the biodegradation capability of chitin *in vivo* and other properties such as solubility, gelling and reactivity. For example, the amount of -acetyl groups can hinder the gel formation of N-acetyl-chitosan.[68] Therefore, the preparation of the appropriate level of acetylated chitin is an important aspect in chitin chemistry. Research has shown that not only is the amount important, the distribution of the N-acetyl groups in the biopolymer chainaffets biodegradation and is discussed here.

Isolated chitin has a high level of acetylation normally deacetylated to prepare chitosan. In so doing, it is important to determine conditions to obtain the desired degree of deacetylation. The chemical preparation of chitosan involves the heating of chitin in a concentrated NaOH solution near boiling temperatures. The efficiency usually leads to 80% deacetylation with accompanying depolymerization. Therefore, one of the challenges has been to prepare highly deacetylated chitosan up to 100% with minimal side effects. Mima et al showed that this was possible by treating chitosan with several repeat treatments of concentrated NaOH at elevated temperatures.[69] The secret to achieving high deacetylation appeared to be shorter exposure times, washing the chitosan with water in between NaOH exposure. Domard et al have also prepared fully deacetylated chitosan starting with 80% deacetylated-chitosan in acetic acid followed by exposure to NaOH.[70]

The distribution of acetyl groups on the chitin backbone has influence on the biodegradation of chitin by enzymes. Aiba demonstrated the presence of random and block copolymer structures in partially N-acetylated-chitosans (PAC) by N-acetylating highly deacetylated-chitosan under homogeneous conditions and moderately deacetylated-chitosan (MAC) under heterogeneous conditions.[71] Based on the characteristics of the resulting PAC/MAC, Aiba concluded that acetylation under homogeneous gave a random distribution of N-acetyl groups while heterogeneous conditions gave a block distribution of N-acetyl groups. Tokura et al, establishing random N-acetyl group distribution of partially deacetylated-chitins prepared under homogeneous conditions, independently corroborated this conclusion.[72] Therefore, heterogeneous deacetylation of chitin gives a block-type distribution of N-acetyl-groups preferring amorphous regions of the biopolymer. There is a finite upper limit in heterogeneous deacetylation, with associated crystallinity change and solubility. Homogeneous deacetylation of chitin gives a random distribution of N-acetyl-groups with an accompanying increased solubility and reduced crystallinity.[73]

The preparation of acetylated-chitosan likewise has received attention. Kurita et al found that the method of was important for effective acetylation.[74] Starting with 90% deacetylated-chitosan, acetylation with acetic anhydride using various solvent (aqueous acetic acid/methanol/pyridine) mixtures to produce several N-acetyl-chitosans. The best water-soluble chitosan was obtained from the preparation of chitosan swollen in a high pyridine containing solvent mixture. The acetyl content was approximately 50% where the distribution of was random.

7.9 OLIGOMERIZATION BY CHEMICAL AND ENZYMATIC MEANS

There has been an increasing desire to utilize chitin and chitosan of more uniform size particularly as oligomers. The benefits of oligomers are readily water-solubility and low molecular weight compounds. Oligomers can be produced by chemical and enzymatic means. Chemical methods involve the breakdown of the biopolymer chain by alkali and concentrated acids that can be slow and yielding oligomers with a wide distribution range. Hirano et al first demonstrated generation of chito-oligosaccharides in good yields with nitrous acid.[75] Later, Allan and Peyron advocated the use of nitrous acid to generate low molecular weight chitosans.[76] They noted that the rate of depolymerization by nitrous acid of chitosan in aqueous HCl was independent of the molecular weight of chitosan and decreased when the degree of deacetylation is high.[77]. Alternatively, Domard and Cartier used the time of incubation to control the oligomer distribution range during oligomer preparation concentrated HCl,. A distribution of D.P. from 2 to 15 could be obtained as distinct isolates with narrow ranged fractions up to D.P. of 37 possible.[78] The effect of crystallinity of β-chitin on the depolymerization characteristics compared to α-chitin has also been studied using concentrated HCl.[79] Naturally, β-chitin was found to be more readily hydrolyzed but the final molecular weight ranges were similar and relatively narrow in distribution. Homogeneous hydrolysis of chitosan in 85% phosphoric acid has been reported in the preparation of oligosaccharides of chitosan of D.P of 16.8 that is water insoluble and 7.3 that is water-soluble.[80]

The use of enzymes automatically imposes mild reaction conditions and should permit better control of the oligomer distribution because of the selectivity of enzymers. One of the earliest reports by Uchida et al was the straightforward incubation of chitosan solution in chitosanase over 1-8 hours to give high and low molecular weight chitosan oligomers.[81] Beaulieu et al reported the use of a reliable source of chitosanase to produce chitosan oligomers.[82] Terbojevich et al showed that the use of papain, an abundant enzyme compared to chitosanase, readily hydrolyzes chitosan with a higher degree of acetylation.[83] A narrow range tetramer to heptamer chitooligosaccharide from chitosan has been shown to be possible using a cocktail of celluase, α-amylase and proteinase.[84] These demonstrations show the utility of the enzyme method.

Recently, the use of laser radiation to depolymerize chitosan has also been reported as an alternative that does not require chemicals. Intense femtosecond laser pulses were found to be effective in fragmenting chitosan in aqueous acetic acid without affecting the degree of acetylation.[85] This again would be a mild means of producing oligomers without exposure to chemical contaminants, advantageous for biomedical applications.

7.10 CHEMICAL SYNTHESIS

Finally, the chemical synthesis of chitin would be a welcome addition to the extraction of chitin from shellfish and fungal sources. Artificially synthesized chitin should be free from proteins and other contaminants that again are advantageous for biomedical applications. Kobayashi demonstrated the possibility of synthesizing chitin utilizing a mixture of chemical and enzymatic means.[86] A chitin with molecular weight of up to 6×10^4 was obtainable using this process. Kuyama and Ogawa have synthesized chitosan oligomers up to a dodecasaccharide starting with glycobiosyl donors and acceptors.[87] Tokuyasu et al have also demonstrated that chitin moiety from a chitosan tetramer can be obtained by action of chitin acetylase acting as an acetylating agent.[88] Naturally, the chemical synthesis of chitin is only beginning but heralds of a promise that could be an alternative to other methods of chitin production that could have an impact on biomedical applications.

7.11 IMPLICATIONS

The challenge posed by the intractability of chitin unleashed the ingenuity of the chitin scientific community that has as a delightful consequence, increased versatility of chitin. Harsh chemicals and reaction conditions have given way to milder conditions and more controlled reactions.

The first outcome has to be the annihilation of the mental paralysis on the **intractability** of chitin. There is now a vast and extensive literature resource that can be tapped for generating existing and new chitin derivatives. The limited scope of chitin may perhaps be unfounded and should now be capitalized on especially for their biomedical potential. That is to say that should a desired derivative be conceived, there are ready starting points that can be called upon. For biomedical applications, this is a bonus as pharmaceutically active agents can be linked to the C-6 or N-2 sites for controlled release.

Second, the 5% LiCl/DMAc solvent system will become the method of choice for chemical reactions of chitin. However, the alkali-chitin and tosyl-chitin routes that have given mixed results at best will remain to be important for some time to come as precursors for some chitin derivatives, until the 5% LiCl/DMAc solvent system broaden its range of derivatives offering.

The preparation of chitin-oligomers will become more refined the molecular weight ranges becoming well defined and reproducibly produced.

The chemical synthesis of chitin is also likely to increase as this is the best means of obtaining the purest and contaminant free source of chitin, paramount in biomedical applications.

In a nutshell, much has been done and it is important to take stock of this activity, and settle on what is needed to move chitin forward. The emphasis should continue to be with α-chitin even though β-chitin has been demonstrated as an easier form to work with. This is because the limited supply of β-chitin may restrict its widespread utility and will always be questioned.

How should we move on from here? For biomedical applications, the minimum number of synthesis steps the better. This will safeguard purity, lower production costs and pollution. This can be achieved by screening the compounds already known in the literature, the most obvious, O-carboxymethyl-chitin. Its properties include anionic character permitting water swellability to solubility and readily degradable by lysozyme make it suitable for several biomedical applications. Standardization of its chemical production and proper characterization of O-carboxymethyl-chitin would bring this derivative closer for use. The oligomerization, enzymatic processes and chemical synthesis should all be pursued with vigor, as these methods are key for the future exploitation of chitin as a biomaterial.

7.12 REFERENCES

[1]   R.A.A. Muzzarelli, Chitin, Pergamon Press Ltd, Oxford, UK. 1977. 102

[2]   T. Sannan, K. Kurita, Y. Iwakura, Makromoleculare Chemie 176 (1975) 1191

[3]   R., Trujillo, Preparation of carboxymethylchitin. Carbohydrate Research 7 (1968) 483--485

[4]   S. Tokura, N. Nishi, A. Tsutsumi, O. Somorin, Studies on Chitin. VIII. Some properties of water soluble chitin derivatives. Polymer J. 15(6) (1983) 485-489

[5]   A.C.A. Wan, E. Khor, J.M. Wong, G.W. Hastings, Promotion of calcification of carboxymethyl-chitin discs. Biomaterials 17(15) (1996) 1529-1534

[6] S. Tokura, S-i. Nishimura, N. Nishi, Studies on Chitin. IX. Specific binding of calcium ions by carboxymethyl-chitin. Polymer J. 15(8) (1983) 597-602

[7] R.A.A. Muzzarelli, Carboxymethylated chitins and chitosans. Carbohydrate Polymers 8 (1988) 1-21

[8] H. Yamada, T. Imoto, A convenient synthesis of glycolchitin, a substrate of lysozyme. Carbohydrate Research 92 (1981) 160-162; S. Hirano, Water-soluble glycol chitin and carboxymethylchitin. Methods in Enzymology 161 (1988) 408-410

[9] K. Kurita, Y. Koyama, S. Inoue, S-i. Nishimura, ((Diethylamino)ethyl)chitins: Preparation and properties of novel aminated chitin derivatives. Macromolecules 23 (1990) 2865-2869

[10] O. Somorin, N. Nishi, S. Tokura, J. Noguchi, Studies on Chitin. II. Preparation of benzyl and benzoylchitin. Polymer J. 11(5) (1979) 391-396

[11] M. Rinaudo, P.L. Dung, C. Gey, M. Milas, Substituent distribution on O, N-carboxymethylchitosans by ^{1}H and ^{13}C n.m.r. International J. Biological Macromolecules 14 (1992) 122-128

[12] S. Hirano, Y. Ohe, H. Ono, Selective N-acylation of chitosan. Carbohydrate Research 47 (1976) 315-320

[13] S. Hirano, H. Inui, T. Mikami, Y. Ishigami, H. Hisamori, An alkaline chitin solution and some O-alkylchitin derivatives. Agric. Biol. Chem. 55(10) (1991) 2627-2628

[14] S. Hirano, N. Hutadilok, K-i. Hayashi, K. Hirochi, A. Usutani, H. Tachibana, Alkaline chitin and chitin xanthate: Preparation, derivatives, and applications. in Carbohydrates and Carbohydrate Polymers, Analysis, Biotechnology, Modification, Antiviral, Biomedical and Other Applications, M. Yalpani, ed., ATL Press, 1993. 253-264; S. Hirano, A. Usutani, M. Zhang, Chitin xanthate and some xanthate ester derivatives. Carbohydrate Research 256 (1994) 331-336

[15] S. Hirano, K-I. Hayashi, K. Hirochi, Some N-acyl derivatives of O-carboxymethylchitosan. Carbohydrate Research 225 (1992) 175-178

[16] K. Kurita, S. Inoue, S-i. Nishimura, Preparation of soluble chitin derivatives as reactive precursors for controlled modifications: Tosyl- and iodo-chitins. J. Polymer Science: Part A: Polymer Chemistry 29 (1991) 937-939

[17] K. Kurita, H. Yoshino, K. Yokota, M. Ando, S. Inoue, S. Ishii, S-i. Nishimura, Preparation of tosylchitins as precursors for facile chemical modifications of chitin. Macromolecules 25 (1992) 3786-3790

[18] K. Kurita, H. Yoshino, S-i. Nishimura, S. Ishii, Preparation and biodegradability of chitins derivatives having mercapto groups. Carbohydrate Polymers 20 (1993) 239-245

[19] K. Kurita, S. Inoue, K. Yamamura, H. Yoshino, S. Ishii, S-I. Nishimura, Cationic and radical graft copolymerization of styrene onto iodochitin. Macromolecules 25 (1992) 3791-3794; K. Kurita, S. Hashimoto, H. Yoshino, S. Ishii, S-i. Nishimura, Preparation of chitin/polystyrene hybrid materials by efficient graft copolymerization based on mercaptochitin. Macromolecules 29 (1996) 1939-1942

[20] Y. Morita, Y. Sugahara, A. Takahashi, M. Ibonai, Preparation of chitin-p-toluenesulfonate and deoxy(thiocyanato) chitin. European Polymer J. 30 (1994) 1231-1236

[21] N. Nishi, J. Noguchi, S. Tokura, H. Shiota, Studies on Chitin. I. Acetylation of chitin. Polymer J. 11(1) (1979) 27-32

[22] N. Nishi, H. Ohnuma, S-i. Nishimura, O. Somorin, S. Tokura, Studies on Chitin. VII. Preparations of p-substituted benzoylchitins. Polymer J. 14(11) (1982) 919-923

102

23 K. Kaifu, N. Nishi, T. Komai, S. Tokura, O. Somorin, Studies on Chitin. V. Formylation, propionylation and butyrylation of chitin. Polymer J. 13(3) (1981) 241-245

24 N. Nishi, A. Ebina, S-i. Nishimura, A. Tsutsumi, O. Hasegawa, S. Tokura, Highly phosphorylated derivatives of chitin, partially deacetylated chitin and chitosan as new functional polymers: Preparation and characterization. International J. Biological Macromolecules 8 (1986) 311-317

25 L. Szosland, Synthesis of highly substituted butyryl chitin in the presence of perchloric acid. J. Bioactive Compatible Polymers 11 (1996) 61-71

26 K. Kurita, M. Ishiguro, T. Kitajima, Studies on chitin: 17. Introduction of long alkylidene groups and the influence on the properties. International J. Biological Macromolecules 10 (1988) 124-125

27 M. Terbojevich, A. Cosani, C. Carraro, G. Torri, Homogenoeus-phase synthesis of chitin derivatives. in Chitin and chitosan: Sources, chemistry, biochemistry, physical properties and applications, G. Skjåk-Bræk, T. Anthonsen, P. Sanford, eds., Elsevier Applied Science, England, UK, 1989. 407-414

28 M. Vincendon, Chitin carbamates. in Advances in chitin and chitosan, C.J. Brine, P.A. Sanford, J.P. Zikakis, eds., Elsevier Applied Science, England, UK, 1992. 556-564

29 Y. Morita, Y. Sugahara, A. Takahashi, M. Ibonai, Non-catalytic photo-induced graft copolymerization of methyl methacrylate onto O-acetyl-chitin. European Polymer J. 33 (1997) 1505-1509

30 M. Sakamoto, H. Tseng, K-i. Furuhata, Regioselective chlorination of chitin with N-chlorosuccinmide-triphenylphosphine under homogeneous conditions in lithium chloride-N, N'-dimethylacetamide. Carbohydrate Research 265 (1994) 271-280

31 H. Tseng, K-i. Furuhata, M. Sakamoto, Bromination of regenerated chitin with N-bromosuccinmide and triphenylphosphine under homogeneous conditions in lithium bromide-N, N'-dimethylacetamide. Carbohydrate Research 270 (1995) 149-161

32 H. Tseng, K. Takechi, K-i. Furuhata, Chlorination of chitin with sulfuryl chloride under homogeneous conditions. Carbohydrate Polymers 33 (1997) 13-18

33 K.S. Chow, E. Khor, New fluorinated chitin derivatives: synthesis, characaterization and cytotoxicity assessment. Carbohydrate Polymers, Accepted for publication 2001.

34 K. Kurita, M. Hirakawa, Y. Nishiyama, Silylated chitin: A new organosoluble precursor for facile modifications and film casting. Chemistry Letters (1999) 771-772

35 K. Kurita, K. Tomita, T. Tada, S-i. Nishimura, K. Shimoda, Squid chitin as a potential alternative chitin source: Deacetylation behavior and characteristic properties. J. Polymer Science: Part A: Polymer Chemistry 31 (1993) 485-491

36 K. Kurita, S. Ishii, K. Tomita, S-i. Nishimura, K. Shimoda, Reactivity characteristic of squid β-chitin as compared with those of shrimp chitin: High potentials of squid chitin as a starting material for facile chemical modifications. J. Polymer Science: Part A: Polymer Chemistry 32 (1994) 1027-1032

37 K. Kurita, Chemistry and application of chitin and chitosan. Polymer Degradation and Stability 59 (1998) 117-120

38 Same as reference 12

39 S. Fujii, H. Kumagai, M. Noda, Preparation of poly(acyl)chitosans. Carbohydrate Research 83 (1980) 389-393

40 K. Kurita, S-I. Nishimura, T. Takeda, N-hydroxyacylation of chitosan with lactones. Polymer J. 22 (1990) 429-434

41 M. Weltrowski, B. Martel, M. Morcellet, Chitosan N-benzyl sulfonate derivatives as sorbents for removal of metal ions in an acidic medium. J. Applied Polymer Science 59 (1996) 647-654; G. Crini, G. Torri, M. Guerrini, M. Morcellet, M. Weltrowski, B. Martel, NMR characterization of N-benzyl sulfonated derivatives of chitosan. Carbohydrate Polymers 33 (1997) 145-151

42 J.V. Gruber, V. Rutar, J. Bandekar, P.N Konish, Synthesis of N-[3'-hydroxy-2',3'-dicarboxy)-ethyl]chitosan: A new, water-soluble chitosan derivative. Macromolecules 28 (1995) 8865-8867

43 R.A.A. Muzzarelli, F. Tanfani, M. Emanuelli, S. Mariotti, N-(Carboxymethylidene)chitosans and N-(carboxymethyl)chitosans: Novel chelating polyampholytes obtained from chitosan glyoxylate. Carbohydrate Research 107 (1982) 199-214

44 R.A.A. Muzzarelli, P. Ilari, M. Petrarulo, Solubility and structure of N-carboxymethylchitosan. International J. Biological Macromolecules 16 (1994) 177-180

45 M. Rinaudo, P.L. Dung, C. Gey, M. Milas, Substituent distribution on O, N-carboxymethylchitosans by ^{1}H and ^{13}C nmr. International J. Biological Macromolecules 14 (1992) 122-128

46 K. Kurita, S-i. Nishimura, S. Ishii, O. Kohgu, T. Munakata, K. Tomita, M. Kobayashi, N-phthaloyl chitosan: A convenient precursor for regioselective modifications to branched and amphiphilic derivatives. in Carbohydrate and Carbohydrate Polymers, Analysis, Biotechnology, Modification, Antiviral, Biomedical and Other Applications, M. Yalpani, ed., ATL Press, 1993. 218-226

47 K. Kurita, K. Tomita, T. Tada, S-i. Nishimura, S. Ishii, Reactivity characteristics of a new form of chitosan. Facile N-phthaloyl of chitosan prepared from squid β-chitin for effective solubilization. Polymer J. 30(4) (1993) 429-433

48 S-i. Nishimura, H. Kai, K. Shinada, T. Yoshida, S. Tokura, K. Kurita, H. Nakashima, N. Yamamoto, T. Uryu, Regioselective synthesis of sulfated polysaccharides: Specific anti-HIV-1 activity of novel chitin sulfates. Carbohydrate Research 306 (1998) 427-433

49 N. Kubota, Y. Kikuchi, Preparation and properties of macromolecular complexes consisting of chitosan derivatives and potassium metaphosphate. Makromolekulare Chemie 193 (1992) 559-566

50 A. Denuziere, D. Ferrier, A. Domard, Chitosan-chondroitin sulfate and chitosan-hyaluronate polyelectrolyte complexes. Physico-chemical aspects. Carbohydrate Polymers 29 (1996) 317-323

51 O.C. Agboh, Y. Qin, Chitin and chitosan fibers. Polymers for Advanced Technologies, 8 (1997) 355-365

52 S. Tokura, N. Nishi, J. Noguchi, Studies on Chitin. III. Preparation of chitin fibers. Polymer J. 11(10) (1979) 781-786

53 S. Tokura, N. Nishi, O. Somorin, J. Noguchi, Studies on Chitin. IV. Preparation of acetylchitin fibers. Polymer J. 12(10) (1980) 695-700

54 Same as reference 51

55 Y.C. Wei, S.M. Hudson, J.M. Mayer, D.L. Kaplan, The crosslinking of chitosan fibers. J. Polymer Science. Part A: Polymer Chemistry 30 (1992) 2187-2193

56 S. Hirano, A. Usutani, T. Midorikawa, Novel fibers of N-acylchitosan and its cellulose composite prepared by spinning their aqueous xanthate solutions. Carbohydrate Polymers 33 (1997) 1-4

104

57   S. Hirano, K. Nagamura, M. Zhang, S.K. Kim, B.G. Chung, M. Yoshikawa, T. Midorikawa, Chitosan staple fibers and their chemical modification with some aldehydes. Carbohydrate Polymers 38 (1999) 293-298

58   S. Hirano, N. Matsuda, O. Miura, H. Iwaki, Some N-arylidenechitosan gels. Carbohydrate Research 71 (1979) 339-343

59   G.K. Moore, G.A.F. Roberts, Chitosan gels: 1. Study of reaction variables. Int. J. Biol. Macromol. 1 (1980) 73-77; G.K. Moore, G.A.F. Roberts, Chitosan gels: 2. Mechanism of gelation. International J. Biological Macromolecules 2 (1980) 78-80

60   S. Hirano, R. Yamaguchi, N. Fukui, M. Iwata, Biological gels: The gelation of chitosan and chitin. in Biotechnology and Polymers, C.G. Gebelein, ed., Plenum Press, New York, N.Y., 1991. 181-188

61   S. Hirano, K. Horiuchi, Chitin gels. International J. Biological Macromolecules 11 (1989) 253-254

62   S. Hirano, A. Usutani, Hydrogels of N-acylchitosans and their cellulose composites generated from the aqueous alkaline solutions. International J. Biological Macromolecules 20 (1997) 245-249

63   Y.M. Lee, S.S. Kim, S.H. Kim, Synthesis and properties of poly(ethylene glycol) macromer/β-chitosan hydrogels. J. Materials Science: Materials in Medicine 8 (1997) 537-541; Y.M. Lee, S.S. Kim, Hydrogels of poly(ethylene glycol)-co-poly(lactones) diacrylate macromer and β-chitin. Polymer 38 (1997) 2415-2420

64   G. Paradossi, F. Cavalieri, V. Crsecenzi, ^{1}H nmr relaxation study of a chitosan-cyclodextrin network. Carbohydrate Research 300 (1997) 77-84

65   A.C.A. Wan, E. Khor, G.W. Hastings, Surface carboxymethylation of a chitin hydrogel, J. Bioactive and Compatible Polymers 12(3) (1997) 208-220

66   E. Khor, A.C.A. Wan, C.F. Tee, G.W. Hastings, Reversible water swellable chitin hydrogel. J. Polymer Science Part A: Polymer Chemistry 35 (10) (1997) 2049-2053

67   Nealda L. bte. M. Y., L.Y. Lim, E. Khor, Preparation and characterization of chitin beads as a wound dressing precursor. J. Biomedical Materials Research 54 (2001) 59-68

68   S. Hirano, S. Tsuneyasu, Y. Kondo, Heterogeneous distribution of amino groups in partially N-acetylated derivatives of chitosan. Agricultural Biological Chemistry 45(6) (1981) 1335-1339

69   S. Mima, M. Miya, R. Iwamoto, S. Yoshikawa, Highly deacetylated chitosan and its properties. J. Applied Polymer Science 28 (1983) 1909-1917

70   A. Domard, M. Rinaudo, Preparation and characterization of fully deacetylated chitosan. International J. Biological Macromolecules 11 (1989) 297-302

71   S-i. Aiba, Studies on chitosan: 3. Evidence for the presence of random and block copolymer structures in partially N-acetylated chitosans. International J. Biological Macromolecules 13 (1991) 40-44

72   H. Sashiwa, H. Saimoto, Y. Shigemasa, S. Tokura, N-acetyl group distribution in partially deacetylated chitins prepared under homogeneous conditions. Carbohydrate Research 242 (1993) 167-172

73   Same as reference 5 [Muzzarelli]

74   K. Kurita, M. Kamiya, S-I. Nishimura, Solubilization of a rigid polysaccharide: Controlled partial N-acetylation of chitosan to develop solubility. Carbohydrate Polymers 16 (1991) 83-92

[75] S. Hirano, Y. Kondo, K. Fujii, Preparation of acetylated derivatives of modified chito-oligosaccharides by the depolymerization of partially N-acetylated chitosan with nitrous acid. Carbohydrate Research 144 (1985) 338-341

[76] G.G. Allan, M. Peyron, The kinetics of the depolymerization of chitosan by nitrous acid. in Chitin and chitosan: Sources, chemistry, biochemistry, physical properties and applications, G. Skjåk-Bræk, T. Anthonsen, P. Sanford, eds., Elsevier Applied Science, England, UK, 1989. 443-466

[77] G.G. Allan, M. Peyron, Molecular weight manipulation of chitosan I: Kinetics of depolymerization by nitrous acid. Carbohydrate Research 277 (1995) 257-272

[78] A Domard, N. Cartier, Glucosamine oligomers: 1. Preparation and characterization. Int. J. Biol. Macromol. 11 (1989) 297-302; T. Li, R. Brzezinski, C. Beaulieu, Enzymatic production of chitosan oligomers. Plant Physiology and Biochemistry 33(5) (1995) 599-603

[79] M. Shimojoh, K. Fukushima, K. Kurita, Low-molecular weight chitosan derived from β-chitin: Preparsation, molecular characteristics and aggregation activity. Carbohydrate Polymers 35 (1998) 223-231

[80] M. Hasegawa, A. Isogai, F. Onabe, Preparation of low-molecular-weight chitosan using phosphoric acid. Carbohydrate Polymers 20 (1993) 279-283

[81] Y. Uchida, M. Izume, A. Ohtakara, Preparation of chitosan oligomers with purified chitosanase and its application. in Chitin and chitosan: Sources, chemistry, biochemistry, physical properties and applications, G. Skjåk-Bræk, T. Anthonsen, P. Sanford, eds., Elsevier Applied Science, England, UK, 1989. 373-387

[82] T. Li, R. Brzezinski, C. Beaulieu, Enzymatic production of chitosan oligomers. Plant Physiology and Biochemistry 33(5) (1995) 599-603

[83] M. Terbojevich, A. Cosani, R.A.A. Muzzarelli, Molecular parameters of chitosans depolymerized with the aid of papain. Carbohydrate Polymers 29 (1996) 63-68

[84] H. Zhang, Y. Du, X. Yu, M. Mitsutomi, S-I Aiba, Preparation of chitooligosaccharides from chitosan by a complex enzyme. Carbohydrate Research 320 (1999) 257-260

[85] M.R. Kasaai, J. Arul, S.L. Chin, G. Charlet, The use of intense femtosecond laser pulses for the fragmentation of chitosan. J. Photochemistry and Photobiology A: Chemistry 120 (1999) 201-205

[86] S. Kobayashi, T. Kiyosada, S-i. Shoda, Synthesis of artificial chitin: Irreversible catalytic behavior of a glycosyl hydrolase through a transition state analogue substrate. J. American Chemical Society 118 (1996) 13113-13114

[87] H. Kuyama, T. Ogawa, Synthesis of chitosan oligomers. in Chitin Handbook, R.A.A. Muzarelli, M.G. Peters, ed., Atec Edizioni, Via San Martino, Italy, 1997. 181-189

[88] K. Tokuyasu, H. Ono, M. Mitsutomi, K. Hayashi, Y. Mori, Synthesis of a chitosan tetramer derivative, β-D-GlcNAc-(1→4)-β-D-GlcNAc-(1→4)-β-D-GlcNAc-(1→4)-β-D-GlcN through a partial N-acetylation reaction by chitin deacetylase. Carbohydrate Research 325 (2000) 211-215

CHAPTER 8: THE REGULATORY ROAD TO APPROVAL FOR CHITIN

8.1 THE PATIENT IS FIRST

The biomedical business is a very unique business. As a healthcare industry, it has only one important priority above all, **patient safety**. While it is important to continuously expound the merits of chitin's biomedical promise, all will dissipate into thin air if **patient safety** is not made priority **one**. Chitin and chitosan have been taken for granted to be biocompatible as they have been isolated from natural sources but this impression disregards the stringent evaluation requirement necessary to establish a material safe for use on humans. It only takes a single error in judgment to taint chitin's reputation regardless of how good chitin is. Extensively documented examples include the Dalkon shield episode that did not help the IUD (intrauterine device) contraceptive devices and more famously, the silicone breast implants.   These episodes ignited legal action that threatened to cripple the biomedical industry.   Many producers and suppliers of materials for the medical device industry responded by withdrawing from the market or imposing excessive indemnity clauses before allowing the sale. While legislature such as the "Biomaterials Access Assurance Act" in the United States will alleviate the problem of biomaterials supply for the biomedical industry, there is no better way than but for the chitin community to put forward every reasonable effort necessary in addressing safety issues.[1]

8.2 CHITIN'S REGULATORY STATUS

Today's reality is, no medical device or product can be marketed in most countries without the appropriate regulatory approval. The procedures to secure this approval are known. There are guidelines and standards set by regulatory agencies and enforced by law that manufacturers and raw material producers must comply with.   Furthermore, each new material will have undergone a rigorous evaluation that is repeated every time the supplier, vendor or the production process is changed, even marginally. Ratner et al have elegantly complied an overview of the requirements for this process and the reader is referred to the reference for further information.[2]

As far back as 1984, it was already clear that a "very uniform raw material" form of chitin would be required for biomedical applications.[3]   Much emphasis was placed in the succeeding years to understand the properties and chemistry of chitin and chitosan. At the 5th ICCC (International Chitin Conference) in 1992, reference to the regulatory status of chitin and chitosan were addressed.[4] At that time, chitosan was only allowed as a food additive in animal feed but not as a substance that was "generally recognized as safe (GRAS)". Weiner listed several concerns that had to be addressed in order for chitin to be approved as a food additive and in pharmaceutical uses, advocating petitions to the US pharmacopeia, European pharmacopeia and Japanese pharmacopeia to identify standards for pharmaceutical applications.[5] In a nutshell, the regulatory status for chitin and chitosan, let alone their derivatives, has some ways to go if widespread biomedical applications are to be realized. There are producers who have taken the lead and introduced chitosan salts of various molecular weights and Poly-N-acetyl-glucosamine that meet regulatory standards. Much more can be done for example, the availability of biomedical grade chitin and one of its leading derivatives, O-carboxymethyl-chitin.

What are the hurdles as it were that chitin has to overcome to obtain the coveted regulatory seal of approval? Depending which side of this regulatory fence one sits on, the list can be seemingly endless. Consequently, in approaching a matter of this importance it is taken for granted that regulators and manufacturers are reasonable people acting in the interest of all, patient, science, medicine and business. Based on this premise, a common ground can be established and exceptions being settled on a case-by-case basis. Over the years, meeting the requirements of the United States, centered around the 1976 Medical Device Amendment to the US-FDA regulations and the Medical Device Directive July, 1998 in the EU, these countries being key markets for medical devices, have become the benchmarks to aim for.

8.3 IDENTIFIABLE TARGETS

The way to attain regulatory credibility for chitin is to demonstrate that as raw materials and as components in medical devices, all steps and processes involved to arrive at the finished goods have met the approving authority's requirements. While not exhaustive, this would center on the design, development and testing of the medical device to acceptable levels both in performance as well as safety, manufacturing under a quality system and surveillance of the product after introduction into the market. This can be comfortably looked at from the following viewpoint:

1. Manufacturing in compliance to Current Good Manufacturing Practices (CGMP),
2. Design and prototyping of medical device to ISO 9001 specifications,
3. Safety testing using the ISO 10993 series as a useful platform,
4. And, sterility, as the method of sterilization can affect the biopolymer in a specific manner.

8.3.1 Current Good Manufacturing Practices (CGMP)

Chitin, chitosan, and, or any of their derivatives, are expected to be supplied as biomaterials. This biomaterial will participate as components in medical devices. While separate, both material and device must be produced in a manner that meet Current Good Manufacturing Practice (CGMP) guidelines to guarantee that products are produced in a manner that meet the quality and performance standards expected by regulatory agencies. In CGMP, the manufacturer is obligated to have:

1. Personnel with the appropriate educational background and provided with the relevant training and experience to perform the assigned job.
2. The facilities' design, structures and function must permit proper operations, security, environmental control, separate rooms for various functions such as weighing rooms and storage area.
3. All equipment used must meet the need specified, suitably sited, kept in operational condition, cleaned, maintained and inspected.
4. A quality assurance program and an independent team that is capable of carrying out all duties associated with quality assurance.
5. The production and process controls including packaging and labeling are performed according to the guidelines.
6. All processes have clearly written procedures and all complaints are filed.
7. For medical devices, there is also a requirement for post market surveillance.

The set-up of a facility to meet these requirements entails not only a capital investment to have a qualified operational facility but considerable effort and time to implement and practice both the quality assurance program and operations. All this has to be viewed in light of chitin's potential for initial penetration into the biomedical market and whether the entry can be sustained.

While CGMP does not normally apply to biomaterials but only to finished devices, manufacturers must ensure that the components or raw materials they use meet mandatory standards, by cooperating closely with their suppliers. The production of chitosan salts by Pronova Biomedical based in Norway is a useful example.[6] In their Protosan™ production process, emphasis is placed on the exclusion of endotoxins and heavy metal contaminants using microfiltration followed by ultrafiltration.

Another primary concern in CGMP production is the verification of the quality of chitin material. As outlined in Chapter 6, many methods to characterize chitin materials, such as the determination of the degree of acetylation and molecular weight, are known and used routinely in research. These same methods can be extended to the routine monitoring of chitin production for quality control. The additional requirements are the preparation of protocols that describe in detail the process of acquiring and analyzing data and the calibration of instruments including maintenance cycles. These protocols must be written, adopted and practiced to ensure conformance that should be independent of operator variations. The facilities must also have the proper controls to ensure data integrity. Therefore, it can be concluded that the science and technology exists for the production of chitin and its derivatives according to CGMP.

8.3.2 Medical Device Design and Prototyping

The development of any medical device is based on a need, normally an end-user such as a clinician, nurse or other biomedical practitioner. In the design of a medical device today, regulatory necessity dictates a documented process for the series of steps, including the reasons for all decisions, leading to a prototype. There are now many international Standards that can be recruited, the most common being the ISO 9001 that features a design component or its European equivalent EN46001. For the present purpose, a brief overview of elements in the design and prototyping process is presented below.[7] #

1.   The device to be manufactured is defined.
2.   The design controls that will meet eventual submission requirement of the device are established.
3.   The inputs essential for the proper functioning of the device are identified and described.
4.   The criteria to accept the device are established before commencing on evaluating the device.
5.   Verification that the design met the user's needs and intended use.
6.   Ascertain that risk analysis had been performed.
7.   Design reviews were held throughout the development period.

\#   (This is an abridged version that can be retrieved from the US-FDA website)

These primary elements, placed in the design process, are a comprehensive formula to bestow the best-case scenario possible in the development of a device.

How does this apply to chitin? Taking the fabrication of a Tissue Engineering scaffold for example, what would be the requisite inputs for the scaffold to function? Size, shape, pore-size, stability, cell adherence and storage conditions are some of the boundaries that must be defined. Similarly, criteria to accept the device would include degradation rate, strength of the scaffold and ability to retain included growth factors. It is necessary to think these matters through as the exercise serves to define the strengths and weaknesses of chitin for that particular application. Careful consideration will contribute to establishing chitin as the biomaterial of choice.

8.3.3 Biological Evaluation of Medical Devices

Much has been stated about the "naturalness" of chitin and its presumed biocompatibility and non-harmful qualities, addressed in Chapter 4. It is important to highlight here that there will also be many other ways to use chitin, especially as derivatives, some of which may be borderline toxic and others may have toxic residues inadvertently included during their preparation. These are legitimate concerns that cannot be excluded. In order to safeguard the continued use of chitin and its derivatives, a system must be set in place to demonstrate and maintain their non-toxic features. The CGMP and quality system are two components of the whole picture.

Material and product testing are other elements that must be included in building the profile of chitin as an approved biomaterial. Mechanical and physical properties evaluations such as the tensile strength or hardness of a material are typically application specific. Sources of the many documented procedures and guidelines include the American Society for Testing and Materials (ASTM) and the British Standards Institute (BSI). Biological properties such as toxicity and biocompatibility are the other important evaluations that must be performed. The primary toxicological tests are biological evaluation issues encompassed by the ISO 10993 document dealing specifically with the biological evaluation of medical device summarized in Table 8.1.

The majority of safety tests are primarily biological-based cell culture *in vitro* studies and animal models (or comparative sciences) *in vivo* methods.

In vitro assessment are studies or tests performed outside a living organism, generally comprising cytotoxicity and cytocompatibility. Cytotoxicity is concerned with the acute cytotoxic response of cell cultures towards the biomaterial while cytocompatibility has to do with the effect of the biomaterial on specific cell type function. Cytotoxicity is a preliminary screening method for a biomaterial to evaluate the impact of the biomaterial on common cell function such as cell viability and proliferation, cell protein content and cell membrane lysis. Cell culture tests are quick, highly sensitive, reproducible and comparatively inexpensive.[8] Cell culture tests are predictors of *in vivo* reactivity. *In vitro* cell cultures can reproduce and therefore simulate the environmental conditions necessary to guarantee the viability of the cells. The aim of these tests is to detect leachable toxic components coming from the material that may be released during the *in vivo* application of the biomaterial device. In cell culture tests, such as the direct contact test, a simple morphological examination of biomaterial/tissue interaction where the assay parameter is the cell morphology is performed. Quantitative tests such as the tetrazolium-based colorimetric assay (MTT test) where cell death is measured is now included as results are directly related to the number of viable cells in the culture. The MTT test is rapid and reproducible.[9]

Test	Objective
ISO 10993-10: 1995 Irritation and sensitization	Estimates the irritation and sensitization potential of test materials and their extracts using appropriate site or implant tissue such as the skin and mucous membrane in an animal model and/or human
ISO 10993-5: 1992 Cytotoxicity	Determines the toxic effects on cells caused by materials and/or its extracts that can result in cell growth retardation, inhibition and death
ISO 10993-11: 1993 Acute Systemic Toxicity	Estimates the harmful effects of test materials and/or its extracts in an animal model systemically
ISO 10993-4: 1992 Hemocompatibility	Evaluates the effects by materials on blood such as hemolysis (the degree of red blood cell lysis), thrombosis, plasma proteins, enzymes using blood from animal models
Pyrogenicity	Evaluates the pyrogenic effect of materials or their extracts
ISO 10993-6: 1994 Implantation tests	Evaluates the local toxic effects on living tissue at the gross and microscopic level when sample is implanted into an appropriate site over a predetermined time period
ISO 10993-13: 1998 Degradation products	Identifies and quantifies of the degradation products from polymeric medical devices
ISO 10993-3: 1992 Mutagenicity tests	Evaluates the effect of test materials and/or its extracts on gene mutations, changes in chromosome structure and number and other DNA toxicities

Table 8.1:   Selected procedures for the biological evaluation of medical devices

The human body is a complicated yet harmonized system where all the cells act together. Therefore, what is observed *in vitro* can only be a simplified version of the true situation in a complete organism. The use of animal *in vivo* models simulates better normal, diseased or defected human disease conditions. In addition, the use of small animal models such as rats is relatively inexpensive and permits rapid completion of experiments. The implantation of the device in animal models also permits the evaluation of the host response.

The first reaction of a living complete biological system to a biomaterial immediately after implantation is the adsorption of blood proteins on the biomaterial. There are more than 150 types of proteins present in blood plasma, making the evaluation of the protein biomaterial interaction difficult to define as it depends on the binding potential of each protein.[10] The deposition of the plasma proteins on a biomaterial is followed by platelets adherence and finally by colonization with fibroblasts and smooth muscle cells. Following implantation, the infiltration of leukocytes also occurs, enhancing the acute inflammatory response. Neutrophils also adhere to the protein layer that are eventually replaced by monocytes. Macrophages also migrate to the implant-tissue interface, generating and controlling the inflammatory response, the extent dependant on the chemical composition and surface environment of the implant.

The soft-tissue response to an implanted biomaterial is important for the evaluation of the bioacceptability of the biomaterial. Soft-tissue reactions to implanted biomaterial varies from total toxicity and rejection to bio-acceptability with mild or moderate foreign body giant cell reactions and fibrous encapsulation leading to bio-integration where native tissue grows to the biomaterial surface and into pores that may be present on the biomaterial.[11] The successful biomaterial is to a large extent determined by its chemical composition, biostability and surface. If the implant is non-degradable and if there is a failure of biostability, the implant would be incompatible in the long-term. For degradable materials, the implant should be biologically resorbed without any toxic effects.

The list of tests in Table 8.1 is termed non-clinical laboratory studies. For the submission of all non-clinical laboratory studies for regulatory approval, they must be conducted with compliance to Good Laboratory Practice (GLP). This requires that all studies are conducted with properly defined protocols that are well written in the form of standard operating procedures and to be performed by properly trained personnel under the supervision of a study director. An independent Quality Assurance Unit reviews all aspects of the study during the lifetime of the study. The requirements of GLP are normally more stringent that CGMP and ISO 9001 because the technical skills level required are higher. All procedures requiring Comparative Sciences models must also comply with strict Animal Welfare Act regulations.

The implications of the ISO 10993 series of tests for chitin are clear. Unlike performance tests that are device dependent, the biological evaluation are general tests useful for verifying the non-toxic profile of chitin materials. Once a particular production process has been validated, only periodic surveillance tests are necessary for the continued production, the type of tests and frequency being documented and performed. This will ensure that supplies of chitin products meet the non-toxic requirement whether they are used for biomedical related research or manufacture.

Finally, under this heading, mention must be made of the immunological evaluation issue. The efficiency of present processes to ensure that the exclusion of residual proteins and other extraneous molecules that may cause immunologic responses in the end-user do not remain is seldom stated. Being a material derived from nature, it is important to consider evaluating chitin products in this respect especially for implants. Whether this is truly warranted and/or required will be defined eventually.

8.3.4 Sterility Issues

Materials and devices for medical use must be sterilized to prevent any lethal risk from infection. A material is considered sterile if demonstrated to be free of all living organisms such as bacteria, fungi and virus. There are many documented methods to determine sterility but normally the USP (U.S. Pharmacopoeia), B.P. (British Pharmacopoeia), J.P. (Japanese Pharmacopoeia), or equivalent procedures are followed. This involves soaking the material or device in a culture media and testing the media for microbial presence. For a material or device, the preliminary process involves a validation procedure. This takes into account the sterilizing process that the device undergoes. Once this is determined it serve as a reference for routine sterility checks. Established sterilization methods are steam autoclaving, gamma irradiation, use of ethylene oxide and glutaraldehyde. Less utilized are microwave, hydrogen peroxide (H_2O_2) and electron beam. In sterilizing chitin, all established methods have been studied as to their effects on the biopolymer's chain length that influences its strength.

There are few reports in the literature dealing specifically with sterility studies of chitin, the few focus on chitosan films. Rao and Shrama evaluated the sterilization of chitosan films with steam sterilization (121 °C, 15 lb/in^2, 15 and 30 min), gamma irradiation (2.5 Mrads), ethylene oxide (900 mg/L, 10 Psi pressure) and 2% glutaraldehyde.[12] All samples were found to be sterile regardless of the method of sterilization but gamma irradiation and ethylene oxide treated samples had reduced mechanical strength attributed to chain scission. In a separate study, gamma irradiation was noted to accelerate wound healing.[13] Although no explanation was provided, chain scission was probably involved, providing perhaps monomeric residues that promoted wound healing. γ-Irradiation again was the focus of a paper by Lim et al who verified chain scission using viscosity measurements.[14] The study was also extended to dry heat and saturated steam.[15] After dry heat treatment, samples became more colored and less soluble, the effect intensifying with higher temperatures and longer exposure times and even more pronounced with steam autoclaving. Crosslinking effects on chitosan by the heat treatments were the most likely cause for these observations.

These studies clearly indicate that the type of sterilization method used on chitin materials can affect the materials' properties and would have to be thoroughly evaluated. The effect of sterilization on functional groups would also have to be considered in future when these are used. Packaging, humidity and temperature changes during storage, the presence of residual chemicals all could have bearing on the final conditions of chitin.[16]

8.4 ROLE OF THE SCIENTIST

Basic research conducted in most academic environments is not subject to have in place the elaborate processes and documentation requirements detailed above. However, present trends in academic research and global realities are seeing an increasing emphasis on industrial sponsored research with institutions encouraging the protection of intellectual property. Verification of experimental results for intellectual property requires documentation that are inherent in quality systems practiced according to GLP, CGMP and the ISO Standards. Furthermore, when basic research is conducted under some form of quality system, it could facilitate the fast track to market so important in today's setting. It must also be emphasized that the basic research component could quite appropriately be part of a product life cycle phase that should incorporate the practices outlined in this chapter. In the drive towards "zero tolerance of safety defect" identifying and incorporating basic research at the time of idea conception, may not be a luxury in the not so distant future.[17]

8.5 OUTLOOK

In conclusion, what is necessary to bring chitin from its present state of material to biomaterial has been shown possible. Outstanding issues that require urgent redress, in particular the sterilization issue has also been pointed out. Clearly the science, technology and processes are available to achieve this. The final outcome would be the availability of medical grade chitin products as depicted in Figure 8.1.

The preliminary processes in the collection of shellfish and fungi followed by pretreatments to remove minerals and proteins can be performed as presently practiced, as in most instances facilities and production are already in place. However, it would be better if this phase is also included as a component of the CGMP program. The controlled processing to obtain very pure chitin and other manufacturing such as the production of chemical derivatives and oligomers must be performed under CGMP. In the future, the inclusion of chemically

114

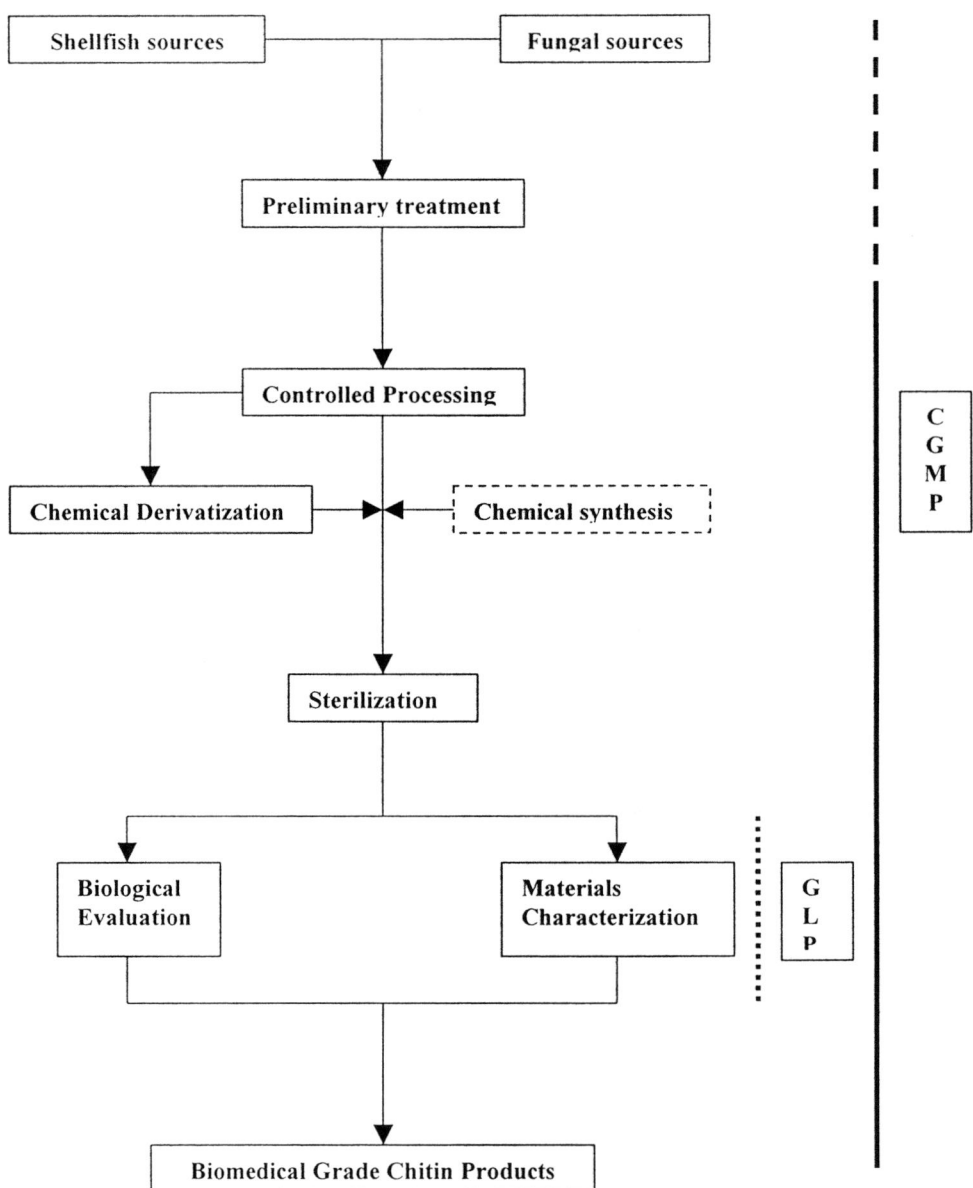

Figure 8.1 Flowchart for the idealized production of biomedical grade chitin products

synthesized chitin materials are likely and are incorporated for completeness. The materials are next sterilized at which point biological evaluations would be necessary for toxicity studies. Simultaneous materials characterization is conducted to define properties such as the molecular weight and degree of acetylation. Obviously, this takes into account that processes to arrive at the various molecular weights and degrees of acetylation were incorporated during the setup validation process. The biomedical grade chitin products from this production sequence can be readily utilized as materials for medical devices and biomedical research.

8.6 REFERENCES

[1]   M. Rouhi, Biomaterials supply remains tight. Chemical and Engineering News 77 (1999) 15-16; P.L. Pabst, K.V. Mills, Supplier liability in medical implants and devices: The "Biomateriasl Access Assurance Act". Tissue Engineering 6 (2000) 189-190

[2]   B.D. Ratner, A.S. Hoffman, F.J. Schoen, J.E. Lemons, Biomaterials Science: An introduction to materials in medicine. Academic Press, San Diego, CA, USA, 1996.

[3]   R.L. Rawls, Prospects brighten for converting chitin wastes to valuable products. Chemical and Engineering News, 62 (1984) 42-45

[4]   J.D. McCurdy, FDA and the use of chitin and chitosan derivatives. in Advances in chitin and chitosan, C.J. Brine, P.A. Sanford, J.P. Zikakis, eds., Elsevier Applied Science, London and New York, 1992. 659-662

[5]   M.L. Weiner, An overview of the regulatory status and of the safety of chitin and chitosan as food and pharmaceutical ingredients. in Advances in chitin and chitosan, C.J. Brine, P.A. Sanford, J.P. Zikakis, eds., Elsevier Applied Science, London and New York, 1992. 663-670

[6]   M. Dornish, A. Hagen, E. Hansson, C. Pecheur, F. Verdier, Ø. Skaugrud, Safety of Protosan™: Ultrapure chitosan salts for biomedical and pharmaceutical use. Advances in Chitin Science, Volume II, A. Domard, G.A.F. Roberts, K.M. Vårum, eds., Jacques Andre Publisher, Lyon 1997. 664-670

[7]   www.fda.gov

[8]   A. Pizzoferrato, G. Ciapetti, S. Stea, E. Cenni, C.R. Arciola, D. Granchi, L. Savarino, Cell culture methods for testing biocompatibility. Clinical Materials 15 (1994) 173-190

[9]   G. Ciapetti, E. Cenni, D. Cavedagna, L. Pratelli and A. Pizzoferrato. Cell culture methods to evaluate the biocompatibility of implant materials. Alternative to Laboratory Animals 20 (1992) 52-60

[10]   M.M. Griffiths, J.J. Langone, M.M. Lightfoote, Biomaterials and granulomas: A companion to Methods in Enzymology 9 (1996) 295-304

[11]   J.M. Morehead, R.G. Holt, Soft-tissue response to synthetic biomaterials. Otolaryngologic Clinics of North America 27(1) (1994) 195-201

[12]   S.B. Rao, C.P. Sharma, Sterilization of chitosan: Implications. J. Biomaterials Applications 10 (1995) 136-143

[13]   R.S. Jayasree. K. Rathinam, C.P. Sharma, Development of artificial skin (template) and influence of different sterilization procedures on wound healing pattern in rabbits and guinea pigs. J. Biomaterials Applications 10 (1995) 144-162

[14]   L-Y. Lim, E. Khor, C-Y. Ling, γ-Irradiation of chitosan. J. Biomedical Materials Research: Applied Biomaterials 43 (1998) 282-290

15   L-Y. Lim, E. Khor, O. Koo, Effects of dry heat and saturated steam on the physical properties of chitosan. J. Biomedical Materials Research: Applied Biomaterials 48 (1999) 111-116

16   H-M. Kam, E. Khor, L-Y. Lim, Storage of partially deacetylated chitosan films. J. Biomedical Materials Research: Applied Biomaterials 48 (1999) 881-888

17   N. Kossovsky, B. Brandegee, Why compliance is not good enough. J. Biomedical Materials Research: Applied Biomaterials 48 (1999) 1-4

CHAPTER 9: TOO LATE THE HERO?

9.1 SITUATION ASSESSMENT

Over the past 25 years, the understanding of biomaterials has witnessed a tremendous growth and the relationship between material characteristics, such as the surface properties of biomaterials and how they interface with cells and tissue, have become identified as important in their interactions with the body. A very positive consequence of this increasing knowledge base is the refinement in the selection and utilization of biomaterials that ensued, out of which came a rekindling of interest in natural materials as biomaterials candidates. The unfortunate outcome for chitin is that it is not alone among the biologically derived biomaterials that have surfaced to meet the increasing demand for better-qualified biomaterials in the 21st Century. While not exhaustive, the list of visible competitors includes hyaluronic acid, collagen and alginates, with chondroitin sulfate and keratan sulfate being minor players and the synthetic analogs, polylactic acid (the leading commercial biodegradable polymer) and polylysine (a key candidate in gene delivery) added for completeness.

How does chitin stack up against the competition in market presence, science and technology, properties and production? It is interesting to note that these materials are competing in the same mix of markets listed in Chapter 2. In most instances, hyaluronic acid and collagen already have products in the market, with the exception being the anticoagulant application, where chitin appears to have the lead (in research) by default that may dissipate in the coming future if hyaluronic acid enters the fray.[1] To obtain a perspective of where chitin stands in light of the competition, it is relevant to examine the matter under the following considerations:

1. What the mission requirements for the biopolymers are and to speculate where/how they may change in time.
2. The suitability of some of chitin's competitors to fulfill these mission requirements.
3. Critically relate and place in perspective, chitin's strengths and weaknesses and chances in light of the competition.
4. The significance of the US-FDA approval process.
5. What are the chances for chitin?

9.2 MISSION REQUIREMENTS

For biologically derived biomaterials, the mission is unquestionably an application that exploits the "naturalness" of the biomaterial. The competitors, hyaluronic acid, collagen, chondroitin sulfate and keratan sulfate, have features common with chitin, mainly, non-toxicity, biodegradability into non-harmful residues that are either assimilated into the body or disposed by the body's normal waste removal channels, good compatibility with cells and tissue, can be processed into shapes and sizes suitable for biomedical applications and sterilized by existing methods. Furthermore, albeit the competitors listed above are obtained from non-human sources, all four biomaterials are ubiquitous to the human body while chitin is the single exception, where only its monomeric constituents are utilized in the human body. Therefore, based on features needed for this class of biomaterials, chitin, at first glance, appears to be at best, on par with its competitors.

From a change perspective, the present landscape of established medical devices dominated by mechanical devices using inert and durable biomaterials, while not likely to disappear, will become progressively less important in the not too distant future as technology trends shift in favor of tissue engineering and minimally invasive procedures to the human body. Tissue engineering is targeted at tissue repair, replacement and regeneration and when successful will have access to a potential market in tissue engineered products worth US$80 billion.[2] The primary role for biodegradable materials in tissue engineering is as scaffolding structures to support initial cell growth and proliferation. There are many strategies reported over the past ten years in the scientific literature using all the key biopolymers, covering the whole spectrum of human parts and organs including cartilage repair, artificial liver and heart valves and peripheral nerve regeneration.

Another role for biodegradable materials would be to act as supports for human stem cells, so distinguished from the normal scaffolding tissue engineered products mentioned above because stem cells are undifferentiated. Stem cells can be motivated to differentiate into all the cell types in the body provided the right stimulus is supplied.

Finally, there is the continuing work on drug delivery and gene delivery requiring biodegradable materials to perform the functions of carriers as well as bioprotectants. Site-specific targeting could also be a primary initiative down the road as the demands for better treatment than can be achieved with present delivery technologies transpire.

Most of tissue engineering and gene delivery technologies still belong at the research level and no stem cell based products have reached the clinical trial stage. When this field explodes in the near future, it will have been made possible in part because specialized biomaterials able to meet the multitude of requirements were developed.

9.3 COMPETITION FROM OTHER BIOPOLYMERS

The key competitors for chitin are hyaluronic acid and collagen, already well established and a survey of their presence and capability is appropriate.

9.3.1 Hyaluronic Acid

Hyaluronic acid (HYA) is a glycosaminoglycan (GAG), an extracellular matrix material present in the body. GAGs are comprised of long polysaccacharide chains linked to a protein moiety.[3] The polysaccharide of HYA comprise of alternating glucuronic acid and N-acetyl-glucosamine. The term hyaluronan embraces the common versions of HYA, the acid form and the sodium salt that is generated at physiological pH. Originally extracted from human umbilical cords and subsequently from the comb of a special strain of rooster, HYA is today obtained biotechnologically, giving HYA that is consistently pure and cost effective to produce. HYA can be crosslinked with many agents such as carbodiimide reagents and epoxides to make them less soluble. HYA can be derivatized at the carboxyl-acid and hydroxyl functionality to sulfates, imparting anticoagulant properties or esterified to impart partial organic solvent solubility for processing purposes. HYA has been combined with synthetic and natural polymers for various applications.

HYA has established extensive credibility as a biomaterial over the past twenty years in ophthalmology, osteoarthritis and wound healing.[4] Hyaluronan containing formulations are claimed to reduce the incidence, extent and severity of postoperative adhesions in the abdomino-pelvic cavity and in pain relieving products that improve joint mobility for osteoarthritis treatments. Hyaluronan are usually modified to enable easier processing into usable forms for use in finished medical devices that take advantage of HYA's properties. For example, the HYAFF® polymers are hyaluronan obtained by esterifying to varying degrees the carboxyl group of the glucuronic acid moiety to give a range of polymers that differ in degradation rate and hydrophobicity. Fibers, films, knitted fabrics, non-wovens, microspheres or sponge-like material forms are known. A number of FDA approved HYA containing products are known and include Healon® (Upjohn-Pharmacia), a viscous injectable gel use in ophthalmology, and Hyalgan® (Sanofi Pharmaceuticals) and SynVisc® (Biomatrix) used in relieving pain and improving joint mobility in osteoarthritis treatment.[5]

In wound healing applications, HYAFF-based polymer products are already in the market in Europe and in clinical trials in the US. Products include membranes having micron dimension perforations used as a delivery system for keratinocytes facilitating migration of epidermal cells to the wound and inclusion of RGD peptide growth factors. HYA has also been shown to enhance peripheral nerve regeneration *in vivo*, with improved nerve conduction velocities and better axon counts (post sacrifice) found in animals given HYA injections.[6] HYA in the form of crosslinked-hydrogels for drug delivery has also been described.[7] The HYA gel could be used for the controlled release of large electronegative molecules. The combination of HYA with other biomaterials is another obvious channel exemplified by a hydroxyapatite-collagen-HYA biomaterial that was reported to have better cohesivity, useful for bone implants.[8] A polypyrrole-HYA material proposed for tissue engineering is another example of a combination material containing HYA.[9] The conducting polymer's role was to provide stimulation to electrically responsive cells with HYA providing the biological compatibility properties. The combination biomaterial was determined to be cell compatible from i*n vitro* studies and *in vivo* results showed vascularization at the implant site indicating angiogenesis promotion by the polypyrrole.

From the foregoing snippets on hyaluronic acid it is obvious that HYA has an established record in its production in pure form, has a defined and suitable, although limited chemistry, and most importantly there are approved HYA containing medical products in the market in the primary segments targeted by chitin. The close structural similarities between chitin and HYA make the extensive chemical derivatization repertoire the only differentiating advantage held by chitin.

9.3.2 Collagen

Collagen is a generic term for a family of extracellular proteins that are essentially polymers of amino acids (polypeptides).[10] At the most basic level, a collagen molecule consists of three chains of polypeptides arranged in a trihelical configuration ending in non-helical carboxyl and amino terminals, one at each end. Collagen can be extracted from a number of animal sources and processed into many forms such as pastes, gels, films, sponges, and felt-like sheets. Crosslinking can be initiated through the pendant amino functionality on lysine residues and carboxyl functionality on glutamic acid residues with glutaraldehyde, epoxies, and carbodiimide reagents and by dye-mediated uv photo-oxidation. Today, collagen is one of the biologically derived biomaterials for the development of medical and other commercial products. The most established commercial collagen-containing devices are wound dressings such as BIOCLUSIVE® transparent dressing, NU-GEL® wound gel, Sof-Foam®, Fibracol®, Comfeel® SEA-CLENS™ wound cleanser, Medfil II® particles and SkinTemp®.

As with HYA, collagen has been investigated for a multitude of similar applications in tissue engineering and drug delivery. Collagen has been utilized as a carrier system of antibiotics for ocular application, and research is on-going as sustained ocular drug delivery systems involving cytokines, 5-fluorouacil and Mitomycin-C. When combined with chitosan, collagen can act as a transdermal delivery system for delivering nifedipine and propanolol hydrochloride.[11] Type I collagen has been demonstrated to self-assemble and found to exhibit the requisite properties such as high wet tensile strength compared to extracted type I fibers.[12] This self-assembly offers the opportunity to prepare collagen scaffolds for tissue engineering applications.

Although collagen is just as established as hyaluronic acid and many products combining HYA or other materials are in the pipeline, there will remain, to varying extents, the concern of the antigenicity of collagen based materials. However, the advent of recombinant human collagen is likely to change this attitude. Therefore, there will always be an opportunity for better biomaterials alternatives to come on the market that may be an opportunity for chitin, provided of course chitin's own protein residue problems are resolved propitiously.

9.3.3 Chondroitin Sulfate

Chondroitin sulfate is another GAG that is usually utilized in a supporting role to HYA and collagen. This is not surprising considering that GAGs in their biological role also support collagen in the extracellular matrix during assembly and function. Chondroitin sulfate has been shown to increase the viscosity of hyaluronic acid at physiological pH, implicating its biological role that may be useful to HYA biomedical applications.[13] Examples of combination with collagen include porous matrices containing chondroitin sulfate useful as tissue engineering applications and in gelatin-chondroitin sulfate gels for the controlled release of cationic antibacterial proteins.[14] It is unlikely that chondroitin sulfate will be a standalone biomaterial and as examples with HYA and gelatin (denatured collagen) show, could also be used in combination with chitin.

9.4 CAN CHITIN COMPETE?

In the brief overview of the competition, chitin appears disadvantageously positioned. In market presence especially wound dressings, hyaluronic acid, alginic acid and collagen-based products abound. The PLA/PGA family of products dominates the biodegradable fixation screws and plates in the orthopedic market. In gene delivery research, chitosan is neck and neck with polylysine. The anticoagulant area is dominated by heparin. Only in blood-contact coating does there appear to be a ray of hope for chitin. In the production of pure biomaterials, again HYA and collagen have an edge. On the topic of science and technology, both HYA and collagen are at least comparable to chitin's base in research and application.

In established products such as wound dressings and orthopedic devices, chitin lags behind. The way forward may be arduous but not insurmountable. Where the playing field is more level is in the emerging technologies of tissue engineering and related fields. Can chitin make a breakthrough?

In any response, the more important consideration is to capitalize on chitin's advantages. The production of medical grade chitin materials must precede any strategy for biomedical applications. As demonstrated by one or two enterprises, it can be accomplished. The major issue that must be addressed is the supply of medical grade chitin materials cost effectively.

More entrants to produce medical grade chitin would favor a reasonable cost and regular supply, encouraging both researchers and medical device producers to seriously consider chitin as a biomaterial. This is the case with hyaluronic acid with 7 major producers throughout the world. It is also timely to look at the isolation of chitin from fungi and diatoms perhaps even biotechnologically and more appropriately its chemical synthesis. This signals an exciting time for chitin.

On the subject of matching the present products in the marketplace, this is a question of producing the better product and finding the right entry point. As the survey of biomedical applications substantiates, there are many ways, even with a given application such as anticoagulant or wound dressing, in which chitin and its derivatives can be applied.

Where chitin may be superior is in its properties and rich chemistry. The processing of various shapes, gels and fibers are known for chitin and its derivatives. In contrast to collagen that denatures above 45°C, chitin is reasonably stable up to 150°C, and chitin can be more tolerant during processing as well as storage. This may have advantage in packaging and handling of chitin products. Perhaps the rich chemistry for chitin may be its ultimate competitive edge. The trend towards combining two or more materials is an indication of the limitations of individual biomaterials. The versatility demonstrated in the chemical derivatization of chitin should extend chitin's usefulness in all the key application areas discussed.

The final comment is the question of chitin or chitosan. There are many ways to address this issue. However, the author has felt this is best left to the individual advocate to champion his or her own cause. Suffice, that for the chitin/chitosan community, the ultimate goal is the success of this biopolymer in biomedical applications.

9.5 SIGNIFICANCE OF THE US-FDA APPROVAL PROCESS

It must be taken for granted that most reputable medical products manufacturers are conscientious in the manner they produce their products, ensuring that the appropriate and necessary standards are met and obtaining approval from their respective country's health authority. With the increasing globalization of trade, certification such as the ISO 9000 series is becoming common. However, where medical products are involved, there is a tendency to be fixated on the US-FDA approval process. While indisputably, the US market is the most important market segment in terms of revenue generation, the reason for the tremendous interest in the US-FDA approval process is the importance placed on recognition. By reputation, it is the most thorough and stringent authority that safeguards the interest of the general public. Many people worldwide perceive a medical product that has obtained a license to market from the US-FDA as having passed the "acid" test. Several products that contain collagen, hyaluronic acid and chondroitin sulfate have obtained US-FDA approval, yet many more products awaiting approval are marketed outside the US, mainly in Europe. It is uplifting to note, that chitin/chitosan containing products are now undergoing this process.

9.6 EPITAPH OR GLORY?

Chitin certainly enjoys a wonderful scientific record that is equal to, if not surpassing its competitors. However where it really counts, in the market place that ensures not only a financial return, but also more importantly, the continuation of chitin science and technology, chitin may have apparently come up short. Chitin's major competitors have proliferated the

122

biomedical market place, while chitin has barely made a dent in the past decade. Is it too late for chitin? There is a common saying used typically to encourage the despondent, "It's not over till the fat lady sings." And, as far as this author is concerned, not by a long shot, as chitin gradually creeps into the very voids left unanswered and unfulfilled by current biomaterials.

9.7 REFERENCES

[1] G. Abatangelo, R. Barbucci, P. Brun, S. Lamponi, Biocompatibility and enzymatic degradation studies on sulfated hyaluronic acid derivatives. Biomaterials 18 (1997) 1411-1415

[2] Obtained from: www.businessweek.com/1998/30/b3588005.htm (1998)

[3] I.F. Radaeva, G.A. Kostina, A.V. Zmievskii, Hyaluronic acid: Biological role, structure, synthesis, isolation, purification and applications (review). Applied Biocehmistry and Microbiology 33 (1997) 111-115; D. Hoekstra, Hyaluronan as a versatile biomaterial for surface treatment of medical devices. Obtained from: www.biocoat.com/hyalvers.pdf (2001)

[4] K.L. Goa, P. Benfield, Hyaluronic acid: A review of its pharmacology and uses as a surgical aid in ophthalmology, and its therapeuticl potential in joint disease and wound healing. Drugs 47 (1994) 536-566

[5] www.xomed.com (2001); V. Listrat, X. Ayral, F. Patarnello, J-P. Bonvarlet, J. Simonnet, B. Amor, M. Dougados, Arthroscopic evaluation of potential structure modifying activity of hyaluronan (Hyalgan®) in osteoarthritis of the knee. Osteroarthritis and Cartilage 5 (1997) 153-160; R.D. Altman, R. Moskowitz, Hyalgan® study group, Intraarticular sodium hyaluronate (Hyalgan®) in the treatment of patients with osteoarthritis of the knee. A randomized clinical trial. J. Rheumatology 25 (1998) 2203-2212

[6] K-K. Wang, I.R. Nemeth, B.R. Seckel, D.P. Chakalis-Haley, D.A. Swann, J-W. Kuo, D.J. Bryan, C.L. Cetrulo, Jr., Hyaluronic acid enhances peripheral nerve regeneration in vivo. Microsurgery 18 (1998) 270-275

[7] R. Tomer, D. Dimitrijevic, A.T. Florence, Electrically controlled release of macromolecules from cross-linked hyaluronic acid hydrogels. J. Controlled Release 33 (1995) 405-413

[8] D. Bakoš, M. Soldán, I.Hernández-Fuentes, Hydroxyapatite-collagen-hyaluronic acid composite. Biomaterials 20 (1999) 191-195

[9] J.H. Collier J.P. Camp, T.W. Hudson, C.E. Schmidt, Synthesis and characterization of polypyrrole-hyaluronic acid composite biomaterials for tissue engineering applications. J. Biomedical Materials Research 50 (2000) 574-584

[10] E. Khor, Method for the treatment of collageneous tissues for bioprostheses (Review). Biomaterials 18 (1997) 95-105

[11] D. Thacharodi, K.P. Rao, Collagen-chitosan composite membranes controlled transdermal delivery of nifedipine and propanolol hydrochloride. International J. Pharmaceutics 134 (1996) 239-241

[12] G.D. Pins, F.H. Silver, A self-assembled collagen scaffold suitable for use in soft and hard tissue replacement. Materials Science and Engineering C3 (1995) 101-107

[13] M. Nishimura, W. Yan, Y. Mukudai, S. Nakamura, K. Nakamasu, M. Kawata, T. Kawamoto, M. Noshiro, T. Hamada, Y. Kato, Role of chondroitin sulfate-hyaluronan

interactions in the viscoelastic properties of extracellular matrices and fluids. Biochimica et Biophysica Acta 1380 (1998) 1-9

[14] J.S. Pieper, A. Oosterhof, P.J. Dijkstra, J.H. Veerkamp, T.H. van Kuppevlet, Preparation and characterization of porous crosslinked collageneous matrices containing bioavailable chondroitin sulfate. Biomaterials 20 (1999) 847-858; A.J. Kuijpers. G.H.M. Engbers, T.K.L. Meyvis, S.S.C. de Smedt, J. Demeesterm J. Krijgsveld, S.A.J. Zaat, J. Dankert, J. Feijen, Combined gelatin-chondroitin sulfate hydrogels for controlled release of cationic antibacterial proteins. Macromolecules 33 (2000) 3705-3713

|  | Doc No.: LB-001 |
| | Revision: A |

| | Page 1 of 2 pages |

| Department of Chemistry | |
| National University of Singapore | LABORATORY OF DR. EUGENE KHOR |

| **Standard Protocol: Purification of Chitin from Shellfish Sources** |

1. PURPOSE

1.1 To purify chitin from shellfish sources.

2. SCOPE

2.1 All shellfish derived sources of chitin from commercial vendors or from other shellfish waste processors.

3. RESPONSIBILITY

3.1 All students and staff of Dr. E. Khor's research laboratory.

4. SAFETY

4.1 All acids and bases must be handled in the fume-hood.
4.2 Safety glasses and gloves must be worn at all times.

5. EXPERIMENTAL

5.1 Materials and Apparatus

5.1.1 Raw Chitin
5.1.2 5% Sodium Hydroxide (NaOH)
5.1.3 1M Hydrochloric Acid (HCl)
5.1.4 Deionized Water

5.1.5 Mechanical Stirrer
5.1.6 5L Beaker
5.1.7 Crystallization Dish
5.1.8 1L Measuring Cylinder

	Doc No.: LB-001
	Revision: A
	Page 2 of 2 pages
Department of Chemistry	
National University of Singapore	LABORATORY OF DR. EUGENE KHOR

Standard Protocol: Purification of Chitin from Shellfish Sources

5.2 Purification of Chitin from Shellfish Sources

5.2.1   Weigh 30g of raw chitin and place into a 5L beaker.

5.2.2   Add 4L of 5% (w/v) NaOH solution.

5.2.3   Stir the raw chitin suspended solution with a mechanical stirrer at 200 rpm for 7 days at room temperature. Adjust the stirring speed so that the solution is swirled homogeneously without any spillage or splashing of the solution. It is also advisable to cover the top of the beaker with parafilm to minimize evaporation of NaOH.

5.2.4   After 7 days of stirring, turn off the stirrer and allow the chitin to settle to the bottom of the beaker.

5.2.5   Decant most of the NaOH solution, and wash the chitin mass with deionized water until the washings are pH neutral as indicated by pH paper.

5.2.6   Add the washed chitin into 1M of HCl solution (91.5ml of 36.5% HCl with 800ml deionized water) ensuring that the chitin is immersed in the acid. It is important to ensure that the chitin is almost dried before transferring into the 36.5% HCl to minimize diluting the solution.

5.2.7   Stir the suspended chitin mixture for 1 hour.

5.2.8   Allow the chitin to settle to the bottom again and decant most of the HCl followed by several washes with deionized water until the washings are pH neutral as indicated by pH paper.

5.2.9   Decant the deionized water and place the chitin into a crystallization dish covered with parafilm. Allow to air dry for 1 day.

5.2.10 Finally, place the crystallization dish containing chitin in a vacuum oven at 50°C for 2 days.

5.2.11 Store the vacuum dried chitin in a desiccator, away from light until required.

	Doc No.: LB-012
	Revision: A
	Page 1 of 4 pages
Department of Chemistry National University of Singapore	LABORATORY OF DR. EUGENE KHOR
Standard Protocol:	**Determination of the degree of acetylation of chitin by infrared spectrophotometry**

1. PURPOSE

1.1 To determine the degree of acetylation (DA) of chitin using infrared spectrophotometry.

1.1.1. The degree of acetylation of chitin determines many properties of the biopolymer including reactivity, processing and biodegradability.

1.1.2. This protocol is intended as a simple, rapid, reliable and cost efficient method to determine the degree of acetylation of chitin.

2. SCOPE

2.1 All sources of chitin i.e. from commercial vendors, shellfish isolated, fungi isolated.

3. RESPONSIBILITY

3.1 All students and staff of Dr. E. Khor's research laboratory.

4. SAFETY

4.1 All acids and bases must be handled in the fume-hood.

4.2 Safety glasses and gloves must be worn at all times.

5. EXPERIMENTAL

5.1 Equipemnt

5.1.1. Infrared spectrophotometer (Bio Rad Excalibur FTIR)

	Doc No.: LB-012
	Revision: A
	Page 2 of 4 pages
Department of Chemistry	
National University of Singapore	LABORATORY OF DR. EUGENE KHOR

| **Standard Protocol:** | **Determination of the degree of acetylation of chitin by infrared spectrophotometry** |

5.2 Reagents and Materials

5.2.1 Chitin
5.2.2 KBr powder (for ir pellets)
5.2.3 Oven
5.2.4 Desiccator
5.2.5 Mechanical and manual grinder

5.3 Procedure

5.3.1 Purify the chitin flakes using the Standard Protocol No.: LB-001.
5.3.2 Dry the chitin flakes in a desiccator at room temperature overnight.
5.3.3 Dry 1g of purified chitin flakes in a vacuum oven for 15 minutes at 110°C.
5.3.4 Allow the flakes to cool down.
5.3.5 Grind chitin flakes into a very thin powder (manual or mechanical method).
5.3.6 Prepare the background for the IR spectrum using a KBr pellet (this background has to be prepared daily because it depends on many parameters such as the humidity in the air).
5.3.7 Use a small amount of chitin powder to prepare the sample.
5.3.8 Mix it with the KBr by grinding them together (manual method).
5.3.9 Obtain IR spectra of the sample (using the software "*Bio Rad Merlin 3.0*").
5.3.10 Align baselines as indicated below at (A) and (B) (using the software "*IR Search Master 6.5*"). Parameters for corrective baselines are Deselect "*Use Endpoints*" and select "*Manual point selection*".
5.3.11 Measure the band absorbance at 3450 cm^{-1} (A_{3450}) and 1655 cm^{-1} (A_{1655}) using the zoom function (Graphical measure).
5.3.12 Calculate the percentage N-acetylation (or degree of acetylation) using the equation:

$$\% \text{ N-acetylation} = (A_{1655} / A_{3450}) \times 115$$

5.3.13 At least five samples have to be analyzed for a valid average of the degree of acetylation.

	Doc No.: LB-012
	Revision: A
	Page 3 of 4 pages
Department of Chemistry	
National University of Singapore	LABORATORY OF DR. EUGENE KHOR
Standard Protocol:	**Determination of the degree of acetylation of chitin by infrared spectrophotometry**

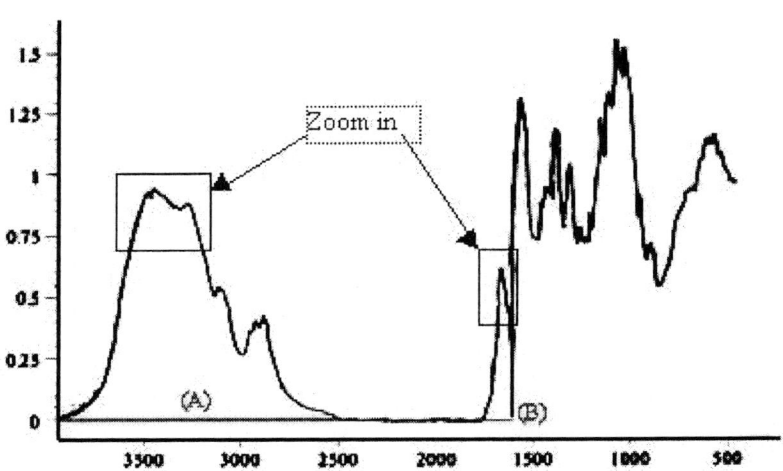

Figure 1:   *IR spectrum in Absorbance after baseline correction.*

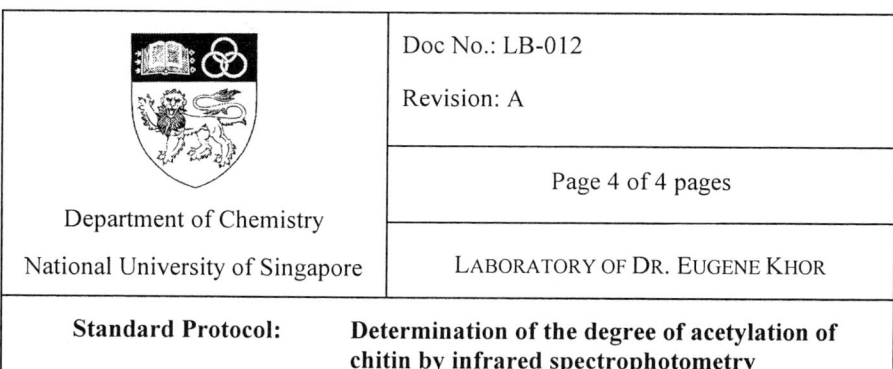

	Doc No.: LB-012
	Revision: A
	Page 4 of 4 pages
Department of Chemistry	
National University of Singapore	LABORATORY OF DR. EUGENE KHOR
Standard Protocol:	**Determination of the degree of acetylation of chitin by infrared spectrophotometry**

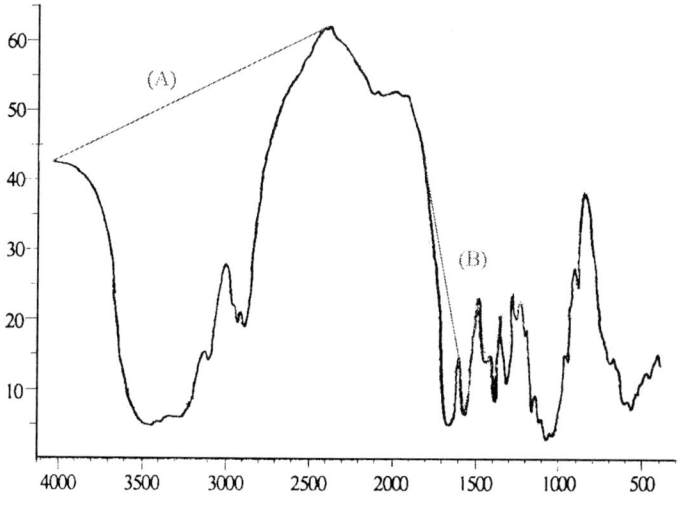

Figure 2:   *IR spectrum with both Baselines (A) and B) in transmittance before baseline correction.*

Note: This protocol is valid for the usual degree of Acetylation of chitin (around 70%). For a different range (as for chitosan) different peaks are used to calculate the DA or use the uv-vis spectrophotometry method (LB-011).

	Doc No.: LB-011
	Revision: A
	Page 1 of 4 pages
Department of Chemistry	
National University of Singapore	LABORATORY OF DR. EUGENE KHOR

Standard Protocol:	**Determination of the degree of deacetylation of chitosan by UV-VIS spectrophotometry**

1.   PURPOSE

1.1 To determine the degree of deacetylation of chitosan using uv-vis spectrophotometry.

1.1.1.   The degree of deacetylation of chitosan defines many properties of chitosan including the reactivity, processing and biodegradability.

1.1.2.   This protocol is intended as a simple, rapid, reliable and cost efficient method to determine the degree of deacetylation of chitosan.

2.   SCOPE

2.1 All shellfish derived sources of chitosan from commercial vendors or from other shellfish processors and fungal chitosan derived in-house.

3.   RESPONSIBILITY

3.1 All students and staff of Dr. E. Khor's research laboratory.

4.   SAFETY

4.1 All acids and bases must be handled in the fume-hood.

4.2 Safety glasses and gloves must be worn at all times.

5.   EXPERIMENTAL

5.1 Equipment

5.1.1.   *Uv-vis spectrophotometer:* The wavelength range should cover the entire visible and near uv with a lower wavelength cutoff of at least 150nm. The slit width should be 1nm or better and the path-length of 10mm as standard for the sample cuvette. The spectrophotometer should have software capable of converting the absorption spectrum into the first derivative spectrum and superimposing spectra.

| | Doc No.: LB-011 |
| | Revision: A |

| | Page 2 of 4 pages |

Department of Chemistry

National University of Singapore | LABORATORY OF DR. EUGENE KHOR |

**Standard Protocol: Determination of the degree of deacetylation of
chitosan by UV-VIS spectrophotometry**

5.2   Reagents and Materials

5.2.1   Acetic acid (Spectra grade).
5.2.2   Only high purity chitosan that is readily soluble in dilute acetic acid shall
be used.
5.2.3   99% minimum N-acetyl-D-glucosamine.
5.2.4   99% D-glucosamine hydrochloride.
5.2.5   Distilled water.
5.2.6   Glass cuvette for uv-vis spectrophotometers with a pathlength of 10mm.
5.2.7   100 ml volumetric flask.
5.2.8   10 ml volumetric flask.

5.3   Procedure

5.3.1   A 100ml (volumetric) sample solution is obtained by dissolving 1mg of
freeze-dried chitosan sample in 10ml of 0.1 M acetic acid and
subsequently topped up to 100ml with distilled water.

Zero Point Crossing (ZPC)

5.3.2   The first derivative spectra of 0.010, 0.020 and 0.030 M of acetic acid
solutions in the range of 190nm to 250nm. Fix the zero crossing point
(ZPC) at 203 nm.

Calibration curve

5.3.3   Prepare reference solutions of 0.005, 0.010, 0.015, 0.020, 0.025, 0.030,
0.035, 0.040, 0.045 and 0.050 mg of GlcNAc per ml of 0.01 M acetic
acid.
5.3.4   Obtain the first derivative spectra of these solutions in the range of
190nm to 250nm.

 Department of Chemistry National University of Singapore	Doc No.: LB-011 Revision: A
	Page 3 of 4 pages
	LABORATORY OF DR. EUGENE KHOR

Standard Protocol: **Determination of the degree of deacetylation of chitosan by UV-VIS spectrophotometry**

5.3.5   Superimpose all 10 spectra and determine the vertical distance from the ZPC to each reference solution, H_1 to H_{10} in mm.

5.3.6   Plot the H_1 to H_{10} values against their corresponding concentrations to obtain a linear calibration curve.

Sample solutions

5.3.7   Obtain the first derivative spectra of sample solutions in the range of 190nm to 250nm. Determine the H value for each sample as outlined in *Calibration Curve*. This is the contribution of GlcNAc to the chitosan sample. Perform five replicates and calculate the average for each sample.

Determination of the DD of chitosan

5.3.8   Determine the DD of the samples by the formula, DD = 100 - [A/(W - 204 A)/161 + A] x 100), Where: A is the amount of GlcNAc determined/204, and W is the mass of chitosan sample used.

Correction of the effect of GlcN on H values

5.3.9   The presence of GlcN will give rise to a larger H value for GlcNAc than expected. Therefore, obtain a reference curve for correcting this discrepancy, as follows:

5.3.10   Prepare a stock 0.10 mg GlcNAc/ml of 0.01 M acetic acid solution.

5.3.11   To 5 x 10 ml of the stock solution, add separately 99, 19, 9, 5.67 and 4 mg of GlcN to give a 1%, 5%, 10%, 15% and 20% GlcNAc solutions (w/w), respectively. Make similarly other solutions of intermediate GlcNAc (%) concentrations, if deemed necessary

5.3.12   Obtain the first derivative spectra for all these solutions.

5.3.13   Measure the H values of the pure GlcNAc solution, H_1, and the H values of the solutions of different percentages of GlcNAc, H_2.

5.3.14   Plot the reference curve H_1/H_2 versus the corresponding GlcNAc percentage

5.3.15   Apply the correction to the H value obtained in 5.3.7 to obtain the final value for GlcNAc for the DD calculation in 5.3.8.

	Doc No.: LB-011
	Revision: A
	Page 4 of 4 pages
Department of Chemistry	
National University of Singapore	LABORATORY OF DR. EUGENE KHOR

Standard Protocol:	**Determination of the degree of deacetylation of chitosan by UV-VIS spectrophotometry**

6.   REPORT

6.1 The make of the uv-vis spectrophotometer.

6.2 The source of the N-acetyl-D-glucosamine and D-glucosamine hydrochloride references.

6.3 The first derivative spectra of acetic acid (5.3.2), references for calibration and correction and samples.

6.4 Calibration and correction curves.

6.5 The DD of the sample.

[Note: enter directly into laboratory notebook; spectra stored in data file properly referenced to laboratory notebook.]

7.   REFERENCES

7.1 RAA Muzzarelli, R Rochetti, Determination of the degree of acetylation of chitosan by first derivative ultraviolet spectrophotometry. Carbohydrate Polymers 5 (1985) 461-472.

7.2 SC Tan, E. Khor, TK Tan, SM Wong, The degree of deacetylation of chitosan: Advocating the first derivative UV-spectrophotometry method of determination. Talanta 45 (1998) 713-719.

SUBJECT INDEX